FLIGHT SAFE AERODYNAMICS

FLIGHT SAFETY AERODYNAMICS

2nd EDITION

Aage Roed

Airlife
England

First published in the UK in 1997
by Airlife Publishing Ltd

British Library Cataloguing-in-Publication Data
A catalogue record for this book
is available from the British Library

ISBN 1 85310 921 5

Typeset by Phoenix Typesetting, Ilkley, West Yorkshire
Printed in England by St Edmundsbury Press Ltd, Bury St Edmunds, Suffolk

Airlife Publishing Ltd
101 Longden Road, Shrewsbury, SY3 9EB, England.

Foreword

The purpose of this book is to explain how airflow, performance, handling qualities and airload problems affect flight safety.

The first edition was published in 1970. The present edition has been thoroughly revised to include problems concerning the latest aircraft types.

This book also has a chapter on helicopter aerodynamics covering the most important helicopter safety problems.

I have also taken the liberty to discuss a number of human-factor problems even if these do not directly concern aerodynamics.

Aage Roed

Contents

CHAPTER 1
FLIGHT SAFETY

Review of Safety Development

First World War pilots were sent to the front with very little flight training. They could expect to live a few weeks after entering combat service.

The US Air Mail began flying mail shortly after the war. Their pilots had a life expectancy of three years. Of the first forty pilots hired, thirty-one died in aircraft accidents.

With this poor safety record neither military nor civil flying could grow. Improvement in safety has been a leading factor in aviation development. Engaged management striving for risk reduction, improved aeroplane design, improved ground facilities, stringent aviation regulations and

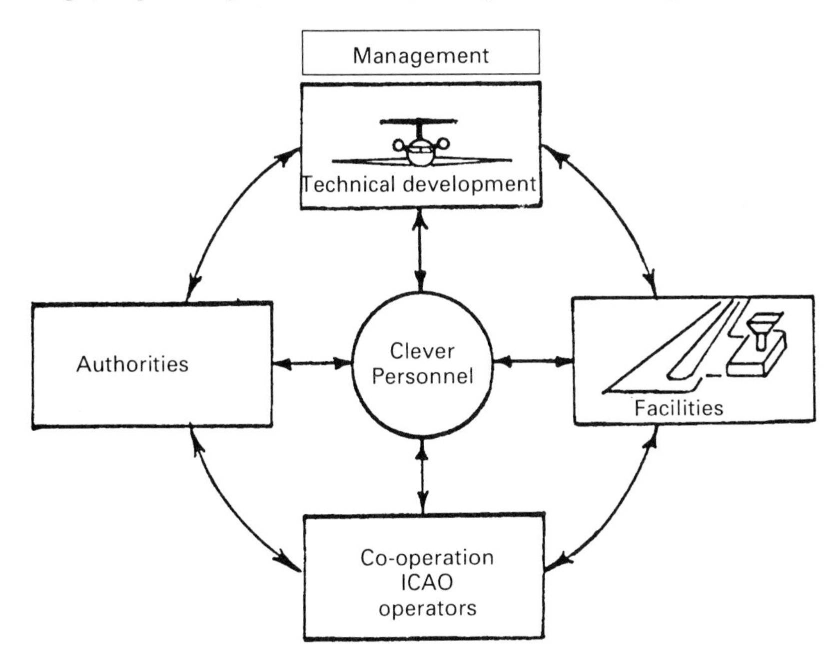

Fig. 1.1 *Factors affecting safety*

international cooperation all have affected each other and created the economic basis for aviation as we see it today. In the centre of the development has been a core of well-trained and dedicated people.

As a result we have seen outstanding safety improvements. An air force which lost about twenty jet fighters a year in combat training some thirty-five years ago today loses one aircraft every two years. During the same period the number of fatalities per 100 million passenger kilometres in scheduled transport aviation has decreased by 98%. A remarkable achievement. The life expectancy of airline crews is now the same as for anyone in a modern society.

Similar safety development is found in all types of aviation. However, as accident rates have gone down, the improvement rates have stagnated. As shown by fig. 1.3 there is no decreasing trend in the number of fatal accidents per 100 000 flight hours during the eleven years from 1982 to 1992.

Low accident rates and stagnated safety improvement does not mean that we have reached a satisfactory safety level. Human suffering and high accident costs motivate continued work for better safety. As an example of

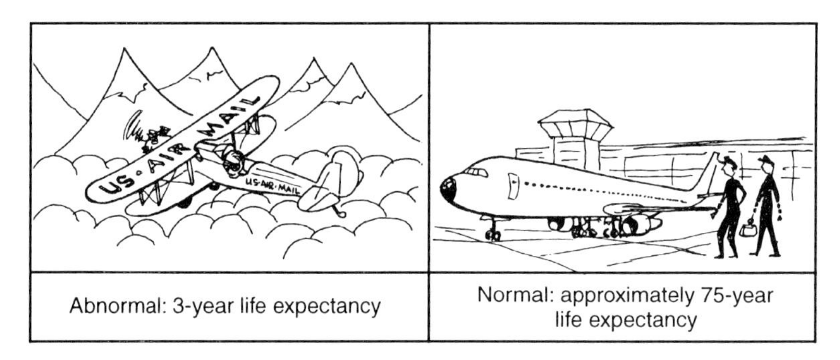

Fig. 1.2 *Life expectancy of aircrews*

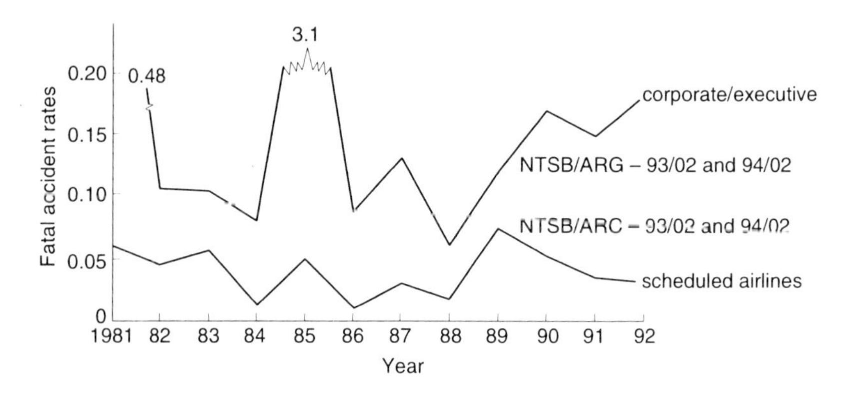

Fig. 1.3 *Fatal accidents per 100 000 flight hours*

what we lose, take a look at the losses in the USA during a ten-year period.

- Airline operations: 248 accidents and 1482 fatalities.
- Regional operators: 1340 accidents and 989 fatalities.
- General aviation: 24 998 accidents and 8850 fatalities.

The worldwide losses are of course many times higher since the accident rates outside the USA are usually worse than the US rates. Most accidents arise from monotonous repetition of trivial causes. Continued improvement is therefore possible.

What Causes Accidents?

Safety risks and accident causes are discussed in detail in other chapters. Here, a general overview of the causes is given. A review of primary airline accident factors is given in fig. 1.4.

The figure shows that crew mistakes dominate the accident picture. However, looking beyond the primary cause we may find that human factors contribute to a majority of the accidents. Improved personnel training does not only reduce crew mistakes; it affects the safety awareness of the whole organisation, can reduce dangerous stress, improve maintenance, teach flight crew how to avoid dangerous situations and in the end reduce costs.

The effect of training on safety is shown in fig. 1.5 where the recent ten-year average fatal accident rates per 100 000 flight hours are given for various types of civil aviation. Professionally trained scheduled airline

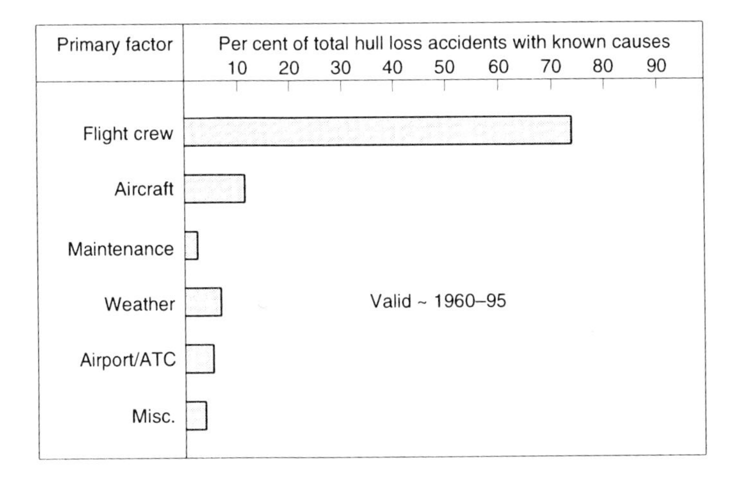

Fig. 1.4 *Statistical review of airline accident causes*

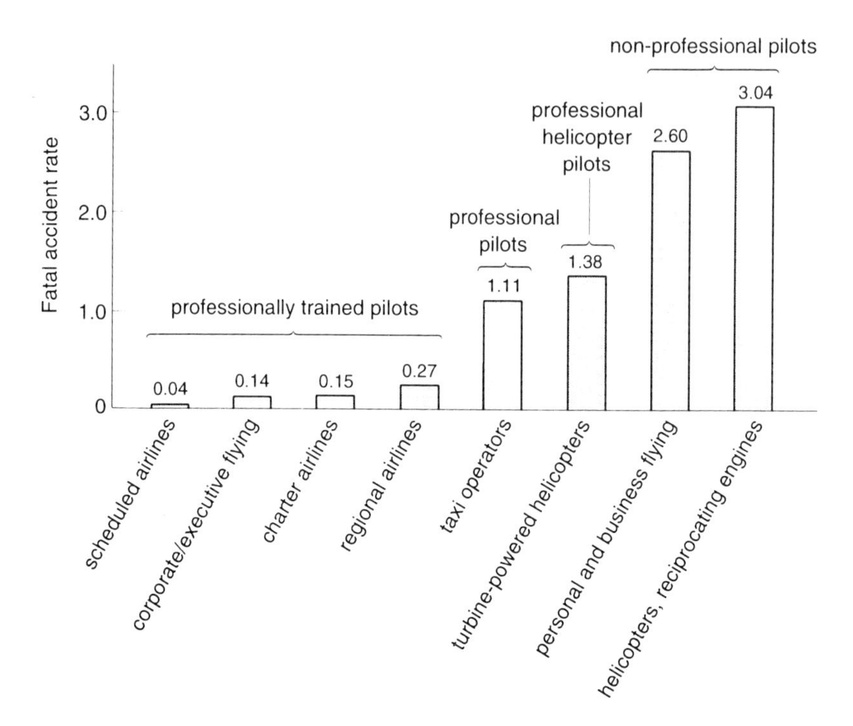

Fig. 1.5 *Fatal accident rates per 100 000 flight hours*

pilots have the lowest rate. The rate for personal and business flying is sixty-five times higher. The comparison is not completely fair because it does not take into consideration the number of take-offs and landings per flight hour. However, comparisons based on the number of departures show the same trend.

The high rate for professional helicopter pilots reflects the higher risks of flying and hovering helicopters close to the ground, compared to long-range fixed-wing flights at high altitudes.

Pilots flying turbine-powered helicopters are often engaged in more difficult operations than those flying light helicopters with reciprocating engines. Still, the piston-engine helicopter accident rate is roughly 120% higher than the rate for turbine helicopters. Heavier helicopters are generally flown by more experienced and better-trained pilots than are those flying smaller helicopters. This surely explains the higher accident rate of pilots of light helicopters . The effects of training and safety management on aircraft accident rates is shown by a survey made by the Boeing Commercial Airplane Division. The investigation showed that of 347 Boeing 737 operators sixty-seven had had accidents but that forty-nine of these had only one accident each. The majority of the accidents were concentrated on a few operators. Of these, two had five accidents each, four had three accidents and twelve had two accidents each. Typical for those with high rates of accidents was:

- poor management control of safety
- poor safety documentation
- poor crew training
- poor cockpit discipline.

Lack of professionalism and lack of training emerge as leading factors causing aircraft accidents.

However, other factors combine to cause accidents. Take a look at the crew cause factors shown in fig. 1.6.

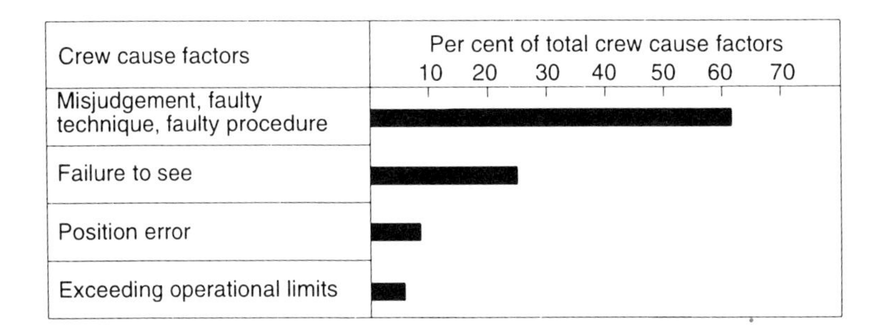

Fig. 1.6 *Review of crew cause factors*

Misjudgement, faulty technique and faulty procedure may be affected by poor cockpit instrumentation, poor warning system design and poorly written flight manuals. It would be nice if we could simplify cockpit design as shown in fig. 1.7.

Instead we tend to overload crews with information which may be difficult to digest in critical situations.

Failure to see is affected by the visibility from the cockpit, by crew seat adjustment problems, outside visibility, sun glare and the closure speed of aircraft which for jets may be as high as $450ms^{-1}$ (approximately 1450 ft/sec). With closure speeds this high there are only a few seconds between conflict detection and head-on collision.

Position errors may be affected by instrument failures. The failure of a cross-bar in a number in a V.O.R./D.M.E. display may change an eight to a zero. There is no warning for this. Navigation data may be affected by changes in computer programming or in failures with poor warning. I.L.S. slopes may be affected by temporary reflections from aircraft or vehicles near the runway ends. I once landed with four over-revved engines when a

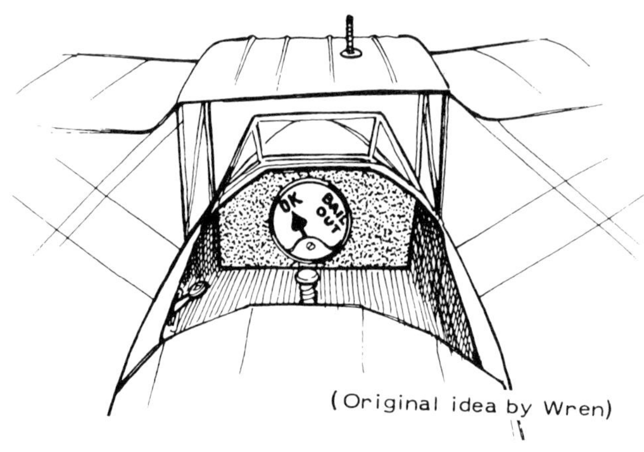

Fig. 1.7

hump in the glide slope told the automatic approach system that the aircraft was far below the glide slope.

A number of factors may affect whether operational limits are exceeded, including the stability and control characteristics of an aircraft, the operational margins and the flight manual's description of the risks. The limits are also affected by factors such as wing contamination, runway contamination and turbulence. Design loads may be exceeded without exceeding the ultimate failure loads. Regularly flying in severe turbulence, frequently at maximum power and with maximum take-off weights, may wear out an aircraft in half the time predicted by the standard 'design use' of the aircraft.

In order to uncover the underlying problems which must be solved to prevent future accidents it is necessary to go far beyond the crew error cause designation. To label an accident 'pilot error' solves nothing.

Accident Prevention

Lack of risk awareness is a major, basic cause of accidents. Awareness requires knowledge. It is too easy to say after having survived a crash that 'I did not know that this was a problem'. However, accidents are repetitions of old mishaps. Hazard information is available for all serious flight problems. All you have to do is to go and get it. Everyone involved in aircraft operations, from managers to individual pilots, should ask themselves these questions:

1. Which accidents have my type of aircraft been involved in? List the accidents and their causes.

2. What are the most common accidents in my type of operational environment? List the causes and compare with the accident list for the aircraft. Any risk of critical aircraft/environment combinations?

3. How do I prepare myself for risk avoidance?

Accident/incident data should be available from data banks, accident/incident reports, manufacturers, and a number of aviation safety publications such as the Flight Safety Foundation publications.

Based on accident prevention analyses it is possible to make simple reminder checklists pointing out the high risks in various flight conditions, such as:

- I.M.C. and night approaches
- winter flying
- thunderstorm avoidance
- checking computer-controlled flight.

Based on identification and analysis of equipment problems, operational problems and human factors it should be possible to develop good accident-prevention training.

The purpose of this book is to point out problems in aerodynamics, handling qualities, performance, and air loads for fixed-wing aircraft and helicopters.

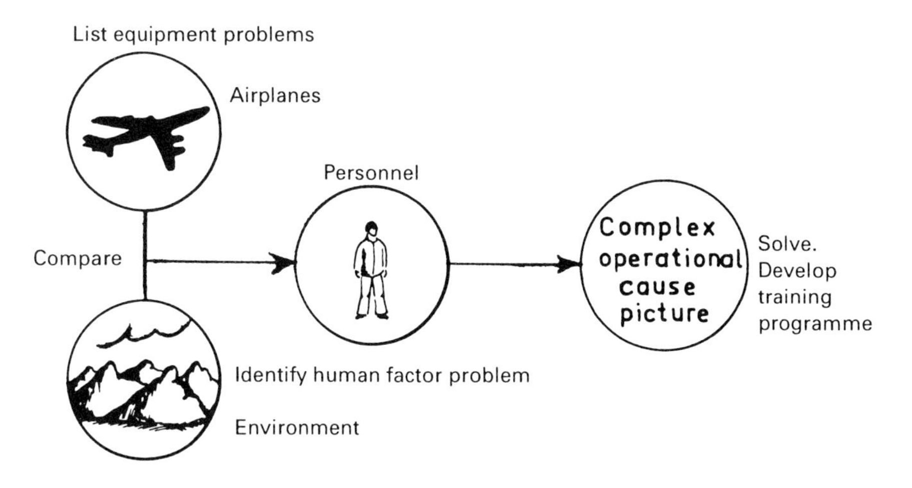

Fig. 1.8 *Accident prevention programme*

Chapter 2
Low-speed Aerodynamics

This chapter gives a review of basic aerodynamics, a subject that is seldom simplified and therefore is considered difficult. Actually it is quite simple.

Note: our reputation may be bad, but the subject is quite simple

Aerodynamic data for aircraft design are obtained from theoretical calculations and wind tunnel tests but must always be verified in full-scale flight tests. Since the flight tests are made with new, factory-smooth aircraft, the reliability of the data may be affected by aircraft ageing and all types of surface contamination.

Aerodynamic data are used for performance calculations, stability and control calculations (handling qualities), and for the determination of air loads (pressure distributions).

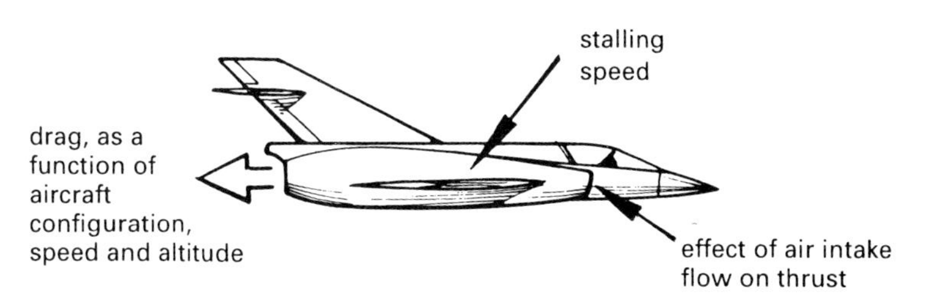

Fig. 2.1 *Performance data*

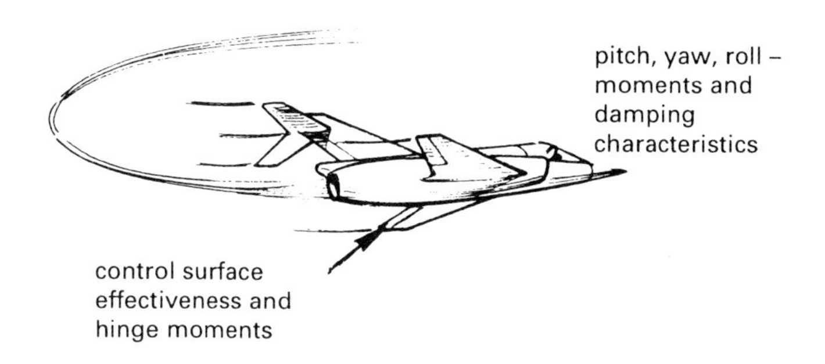

Fig. 2.2 *Handling qualities data*

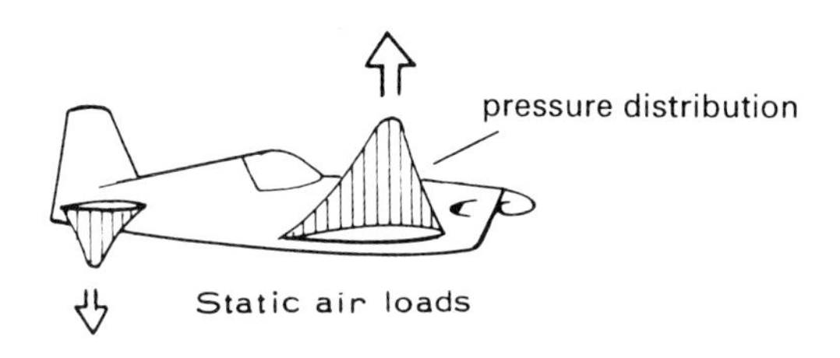

Fig. 2.3 *Static air loads*

Aerodynamic data are divided into two main groups: low-speed data with no compressibility effects; and high-speed data where the compressibility effects are considerable.

Low-speed Frictionless Flow

Why does the pressure of air flowing around a body change? There is a basic law of nature that states: energy cannot appear or disappear by itself, it only changes form.

The ball in fig. 2.4 would bounce forever if it were not for friction. Its total energy would be constant and alternate between potential energy due to its height above the floor and dynamic energy due to gravitational acceleration.

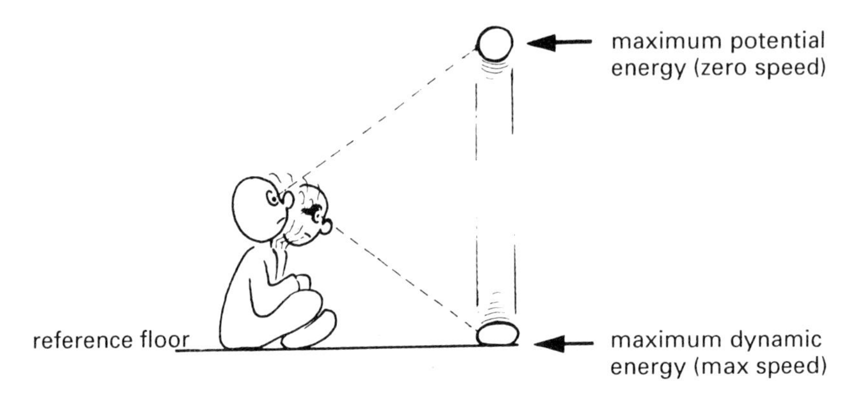

Fig. 2.4 *The sum of energy is constant*

Now, one may assume that the energy of low-speed air consists of two types of pressure: static pressure and dynamic pressure.

For flow in a tube, or flow around a body at constant speed, the sum of static and dynamic pressures are always constant, (fig. 2.6.) Bernoulli's theorem states that the total pressure $P_o = P_s + q$ is constant. It can be shown that the dynamic pressure $q = \frac{1}{2}\rho v^2$ where ρ is air density and v is local flow velocity.

The dynamic pressure increases twice as fast as speed because it is determined by both the speed and the volume of air striking an object. As speed is doubled, twice as much air at twice the speed strikes an object and quadruples the pressure.

This means that as the speed of the air increases in the narrow section of the tube, the dynamic pressure increases and the pressure against the tube walls drops to maintain constant total pressure.

The same holds true for constant speed flow round a cylinder (fig. 2.7). At the points A_1 and A_2 on the cylinder where the flow streamlines part and meet again the local flow velocity must be zero. Therefore the dynamic pressure is zero and the static pressure is at maximum. The pressure at these two 'stagnation points' must be equal and opposite in direction.

At the points B_1 and B_2 the local flow velocity has increased to a maximum and the static pressure against the cylinder wall has reached a minimum.

Thus the air flowing against an object in a frictionless flow is slowed down to a high stagnation pressure at the front stagnation point, thereafter accelerated to low static pressures and again slowed down to a

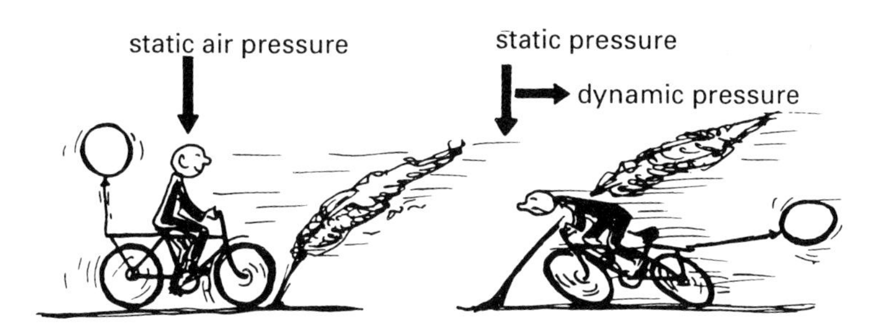

Fig. 2.5 *Static pressure (due to the depth of the atmosphere) and dynamic pressure (due to air motion)*

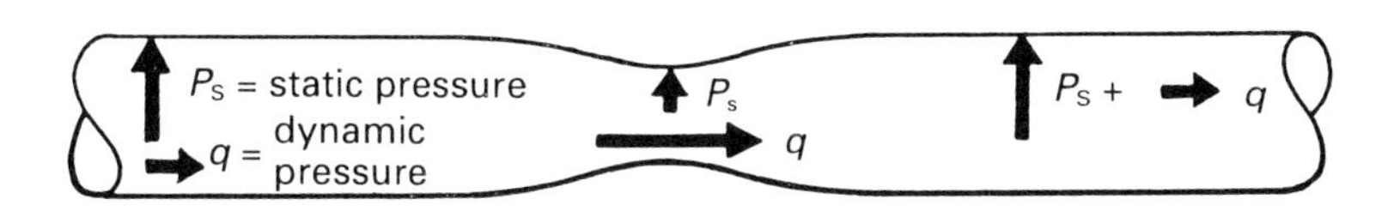

Fig. 2.6 *Flow in a tube*

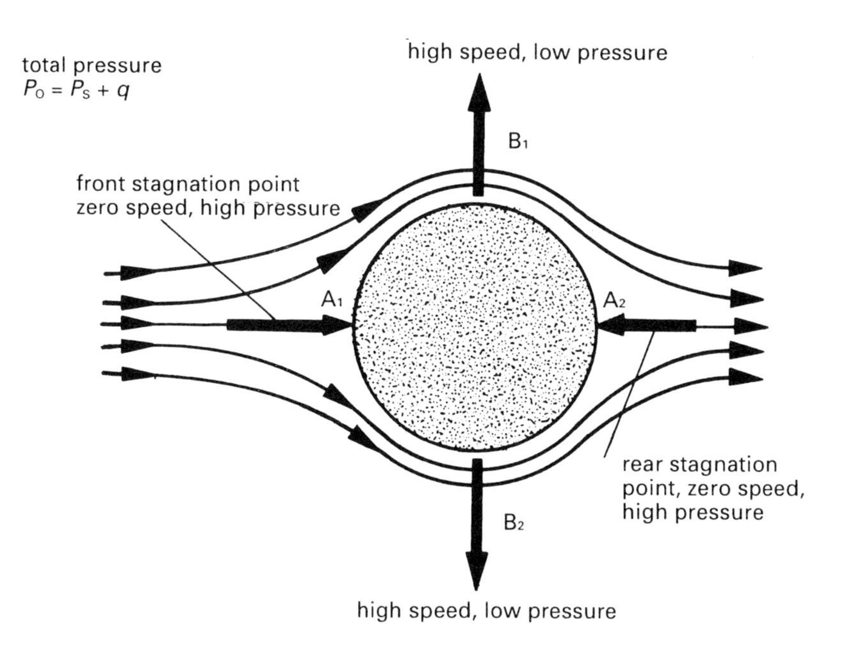

Fig. 2.7 *Flow round cylinder*

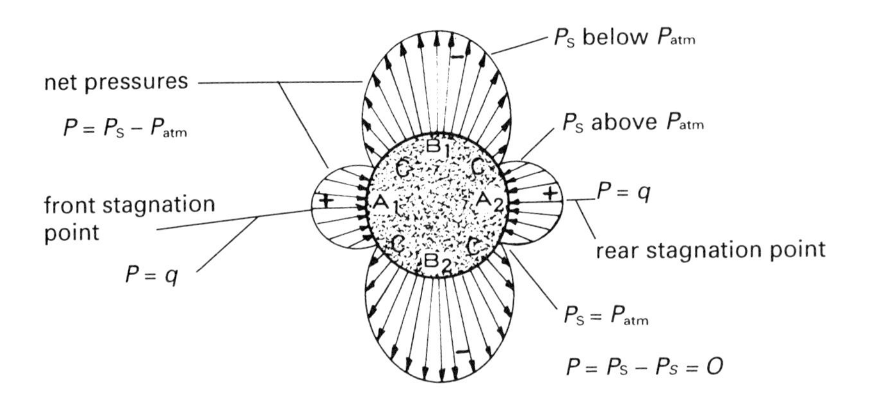

Fig. 2.8 *Net pressure distribution round a cylinder*

high forward-acting stagnation pressure at the rear stagnation point.

In the undisturbed flow approaching the cylinder the static pressure equals the atmospheric pressure (P_{atm}) and the dynamic pressure q is determined by the air density and the free stream velocity (the air speed). The pressure at the front stagnation point therefore equals the atmospheric pressure plus the dynamic pressure. At other points on the cylinder (or any body) where the local flow velocity equals the free stream velocity, the static pressure equals the atmospheric pressure (since the local q value equals the free stream value).

When discussing pressure distributions on bodies (aircraft) the atmospheric pressure is deducted from the total static pressure acting on the surfaces since it is equal in all directions and therefore has no effect on lift or drag. The resulting 'net' pressure distribution ($P_s - P_{atm}$) acting on a cylinder is shown in fig. 2.8.

Due to symmetry, the forces forwards and backwards must be equal and cancel each other. This is also true for the vertical force components and the cylinder will therefore have no drag and no lift.

The atmospheric pressure, and thus the flight altitude, can be measured at any point C. The dynamic pressure, which can be calibrated to give air speed, can be calculated by measuring total pressure at any point, since it is constant, and deducting the atmospheric pressure.

Lift

Lift is a reaction force. When a wing is set at an angle of attack against an airflow it deflects air downwards. As a result, an upward-acting reaction force is obtained (fig. 2.9a).

On the wing profile nose and upper surface of the wing where the air flow is deflected towards the surface a negative centripetal pressure (a suction force) is obtained. It turns the airflow in a curved path along the upper surface. The lowest pressure is obtained at the wing profile nose where the rate of change of flow direction is largest. As the flow approaches the wing trailing edge the pressure increases towards the free stream pressure behind the wing.

On the wing lower surface, near the profile nose, a high stagnation pressure is obtained. This pressure gradually decreases towards the free stream pressure behind the wing.

In the low-pressure area the local flow velocity increases and in the high-pressure area it decreases in order to maintain constant total pressure. Interaction between the local flow velocities and pressures creates the pressure distribution shown in fig. 2.9b. The high pressure below the wing and the low pressure above it cause the airflow to sweep up in front of the wing. The upsweep increases the angular change of flow direction around the profile nose and this increases the suction forces at the wing leading edge.

The lift force resultant of this peaky type of pressure distribution acts along a line located at 25% of the wing chord.

Lift at a constant angle of attack changes with the square of the speed, i.e. it is proportional to the change of dynamic pressure. When the speed is doubled, twice as much air at twice the speed is deflected downwards. This quadruples lift.

The dependence of lift on v^2 also means that as speed is reduced the wing angle of attack must be increased twice as fast as speed is lost. When speed is halved the angle of attack must be increased four times.

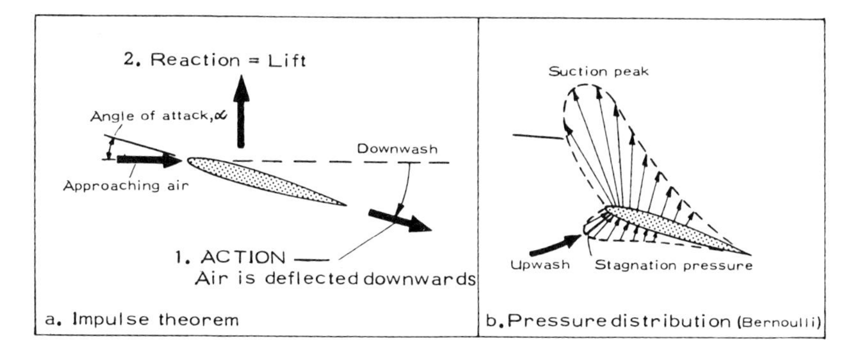

Fig. 2.9 *Lift, a reaction force*

Lift Coefficient, Lift Curve Slope

Lift data are always presented in coefficient form:

$$\text{lift coefficient } C_L = \frac{n_z W}{qs}$$

where n_z is the load factor, W is the aircraft weight, q is the dynamic pressure and s is the wing area.

Lift curve slopes depend on wing aspect ratios, i.e. on the ratio of wing span to mean wing chord (fig. 2.10). Lift is lost near the wing-tips, and as the size of the wing tips relative to the span increases, the losses increase.

The lift curve slope of a soaring aeroplane wing is therefore roughly twice as large as the slope of delta wings and the swept wings of fighters. This affects both angles of attack and drag at take-offs and landings.

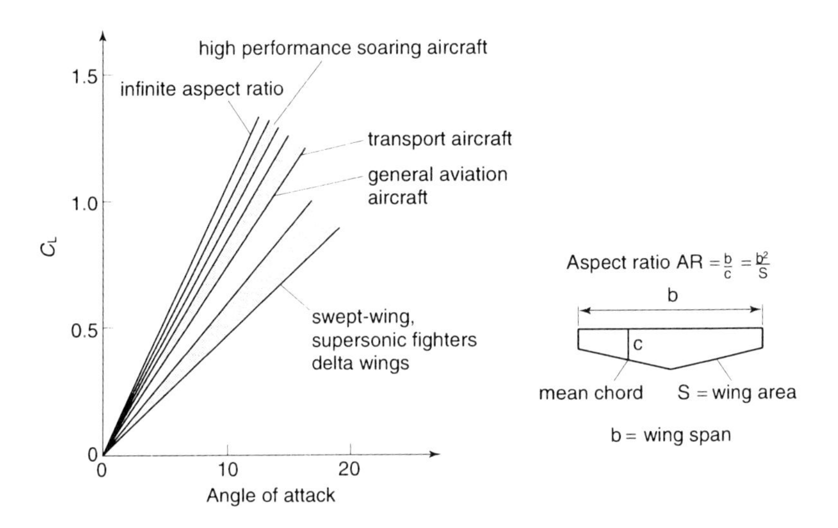

Fig. 2.10 *Lift curve slopes for different aspect ratios*

Friction, The Boundary Layer, Profile Stall

Due to friction a thin boundary layer of slowed-down air is created as air flows along an aircraft surface. The boundary layer may be either laminar (parallel flow) or turbulent (fig. 2.11).

Laminar boundary layers require extremely smooth surfaces. When the flow accelerates from high pressure at the stagnation point A (fig. 2.12) to

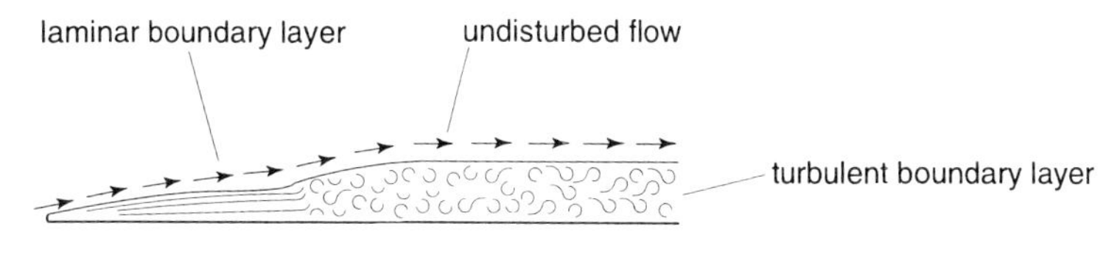

Fig. 2.11 *The boundary layer*

maximum speed at the low pressure point B along the surface of the blunt body, the boundary layer flow is stabilised (accelerated) by the pressure drop. However, behind the suction peak at B, the boundary layer runs into continuously increasing pressure, is slowed down, and finally separates from the surface at some point C. Behind the separation point, a wide, turbulent wake is obtained.

The locations of the point C depend on the smoothness of the surface. If the surface is extremely smooth and the boundary layer is laminar the flow may separate a short distance behind B due to the very low speed in the laminar flow close to the body surface. In turbulent boundary layers the flow speed close to the surface increases, due to mixing effects, and the separation points therefore move aft.

For conventional aircraft wing profiles with turbulent boundary layers, the flow separation point is very close to the trailing edge, at low angles of attack. However, when a certain angle has been reached the mixing becomes insufficient and the flow separation point on the critical side of the profile moves forward, the wing stalls (fig. 2.13).

The location and forward movement of the flow separation point depends on the profile shape. For very thick profiles (blunt profiles) the flow separation may occur somewhat ahead of the trailing edge even at zero angles and the separation point moves forward at fairly low angles of attack. For

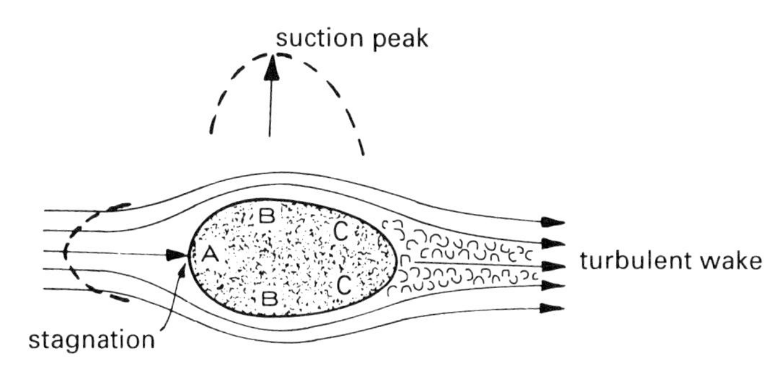

Fig. 2.12 *Flow separation*

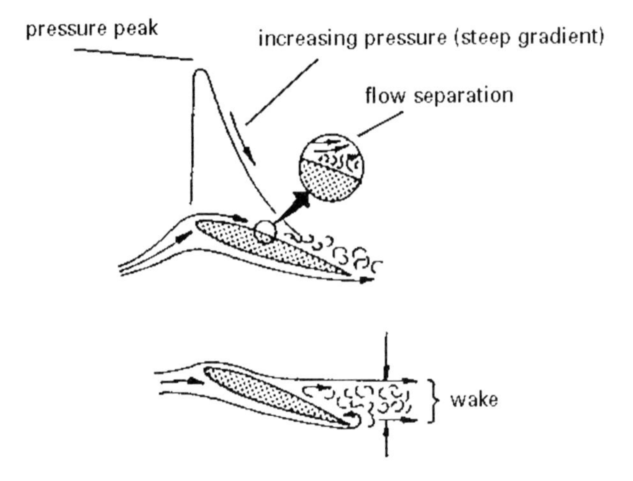

Fig. 2.13 *Profile stall*

profiles in the medium-thick range (thickness ratio 12%–18%) the flow separation creeps forward fairly slowly from the trailing edge and a smooth stall and high maximum lift are obtained. As the profile thickness decreases, the forward motion of the flow separation point becomes very rapid and a sharp, sudden type of stall is obtained (fig. 2.14).

Finally, for very thin profiles (below 6%–7%) flow separation occurs at the leading edge at fairly low angles of attack. In this case the separation does not lead to an immediate stall. A rotating flow (a vortex) is created at the leading edge and the main flow reattaches behind the vortex. As the angle of attack of the profile is increased the vortex grows and finally bursts. A stall-type flow separation is then obtained.

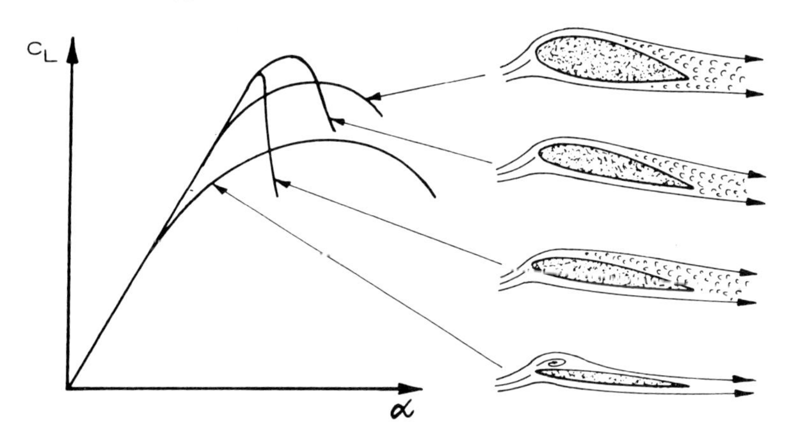

Fig. 2.14 *Typical profile stalls*

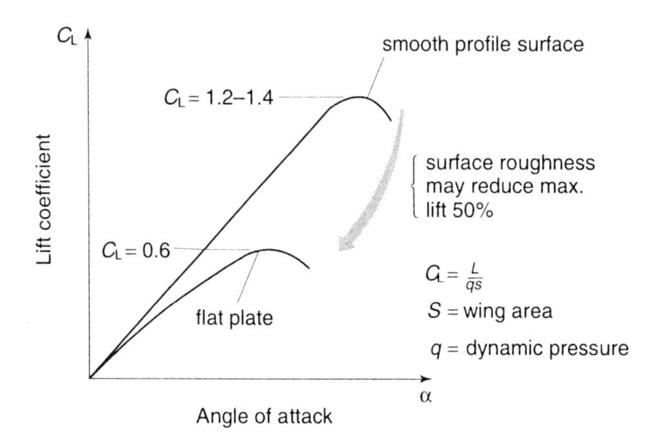

Fig. 2.15 *Effect of surface roughness on maximum lift*

Surface roughness may have a very large effect on the maximum lift coefficient of a wing profile. The maximum coefficient measured on a smooth profile may drop 50% if the surface is worn or contaminated (fig. 2.15; *see* also chapter 10, Contaminated Wings).

Wing Stall

Fig. 2.16 *Stall requirements*

When an aircraft stalls, the flow separation should preferably begin in the wing-root section and spread slowly towards the wing-tips. At stall, the aircraft should pitch down slowly.

The gradually spreading wing-root stall buffets the aircraft before full stall is reached and ensures good aileron control down to stall. When the root section stalls, the downwash of the air behind the wing decreases. This reduces the down-load on the tailplane and the aircraft pitches down to lower angles of attack as the wing stalls. This speeds up stall recovery.

It should be noted that the flow does not reattach to the wing at the original stall angle during recovery. The angle of attack must be reduced quite a few degrees below the stall angle before reattachment is obtained (fig. 2.17).

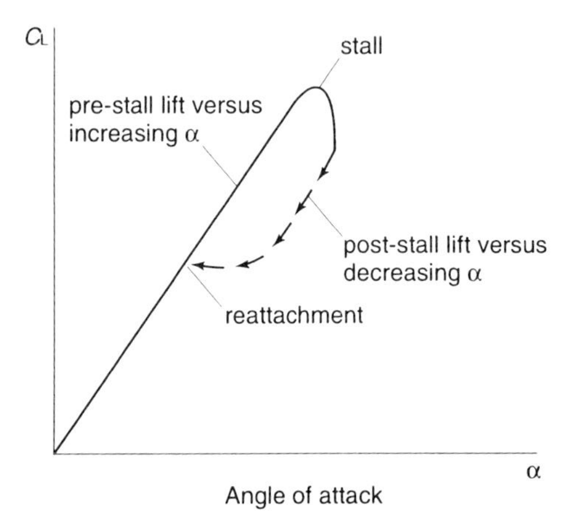

Fig. 2.17 *Hysteresis at stall*

Rectangular wings stall first at the root section and pitch down as the stall develops. Tapered wings with taper ratios below 0.5 ($c_t/c_r < 0.5$ where c_t is the tip chord and c_r is the root chord) have initial stall near the tips and are twisted nose down toward the tips or have drooped profile noses outboard to prevent wing-tip stall. If flight tests show unsatisfactory stall behaviour, stall strips, which reduce the stall angle, may be fitted to the wing leading edges in the root sections (fig. 2.18).

On swept-back wings, the boundary layer turns outboard and flows toward the wing-tips. The reason for this is quite simple (fig. 2.19).

Consider a line normal to the plane of symmetry on a swept-wing aircraft. The pressure decreases from point A towards point B since the line approaches the low-pressure area near the leading edge as it runs in a spanwise direction. The resulting side-force turns the low-energy boundary layer flow in the spanwise direction. As a result, the boundary layer thickness on the outboard sections of the wing increases and an early wing-tip stall may be obtained.

The loss of lift at the wing-tip will, since the tip is located behind the centre of pressure of the aircraft, result in a nose-up pitching moment that may make it difficult to control the aircraft at high angles of attack.

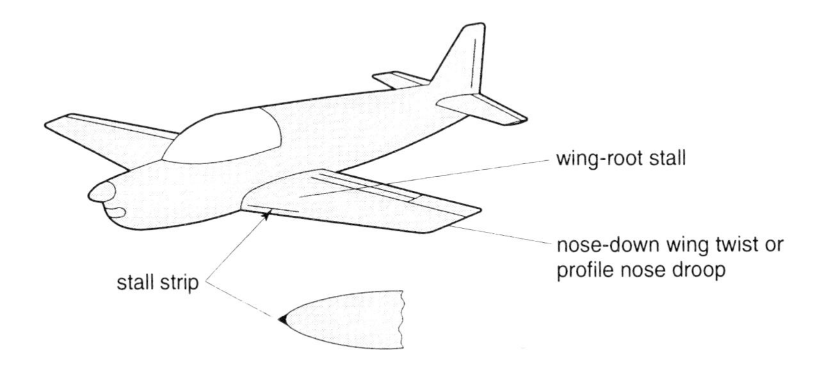

Fig. 2.18 *Factors improving wing stall*

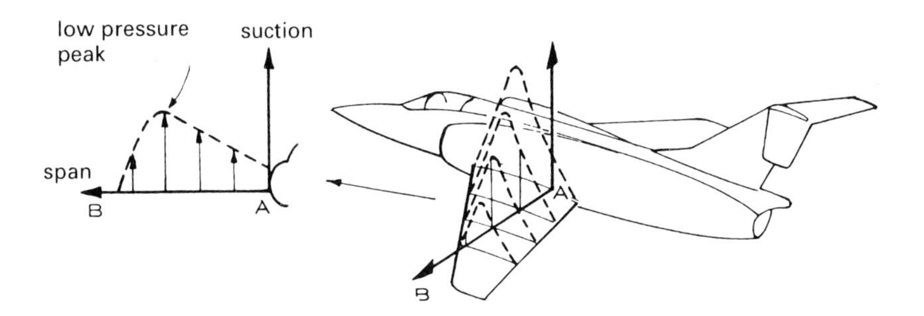

Fig. 2.19 *Spanwise pressure distribution on swept wing*

One way of reducing the effects of spanwise flow is to place a boundary layer fence across the wing, as shown in fig. 2.20, and drain off part of the boundary layer before it reaches the outboard wing sections.

As the wing sweep increases and the relative profile thickness decreases the flow separation starts with a leading edge separation bubble that (if the sweep is sufficiently large) forms a leading edge vortex flow which drains off near the wing-tip (fig. 2.21).

For wing sweeps between 25° and 50° and profiles in the 8% to 10% thickness range, a combination of leading edge vortex flow and spanwise boundary layer flow may be obtained. In this case, care must be taken not to destroy the leading edge vortex flow with the fence used to drain off the spanwise boundary layer flow. The fence must start well behind the leading edge.

As the leading edge sweep approaches 50°–60° the vortex flow becomes dominant and contributes quite a bit to the lift of the wing. The stall picture changes and for a 60° delta wing the following happens (fig. 2.22).

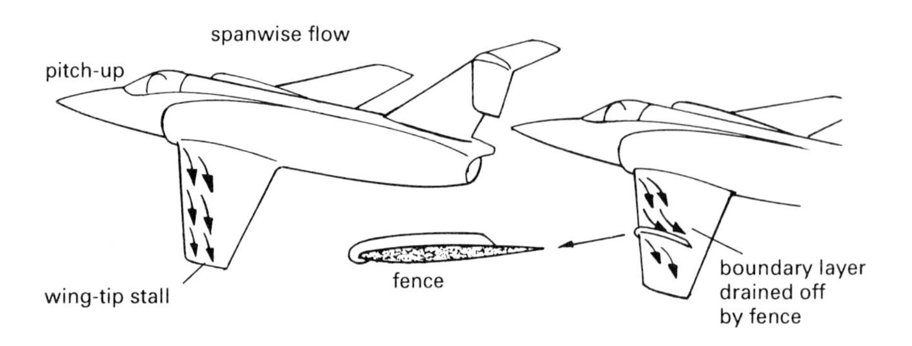

Fig. 2.20 *The boundary layer fence*

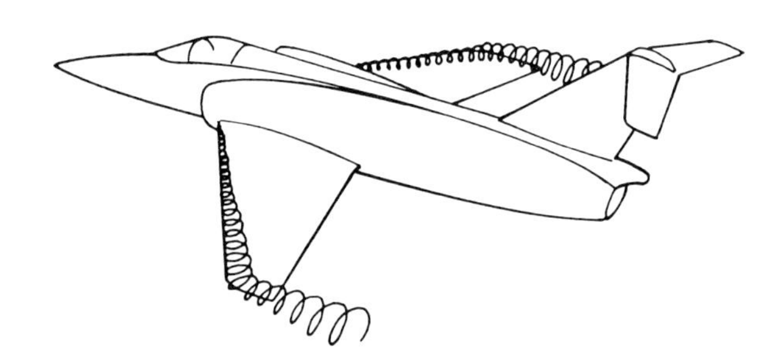

Fig. 2.21 *Leading edge vortex*

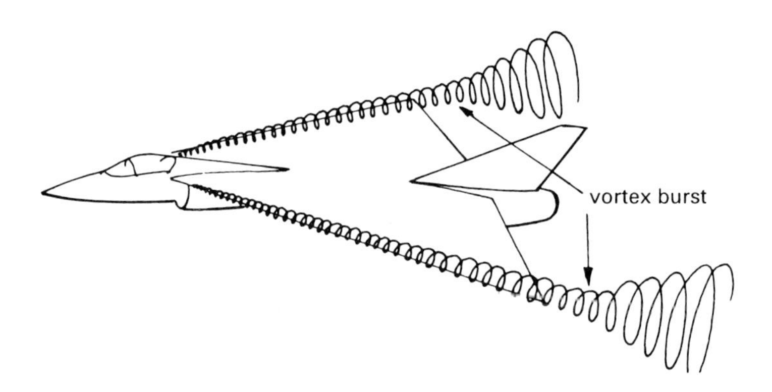

Fig. 2.22 *Delta wing vortex flow*

A fairly stable vortex flow is obtained and flow disturbances can be delayed up to high angles of attack. As the vortex trails from the wing into the air behind the aircraft it flows from an area of low pressure into an area of increasing static pressure. At some distance behind the wing the pressure increase makes the vortex burst and form a funnel-shaped region of slowly rotating 'dead air'. The pressure in the vortex then increases rapidly.

With increasing angle of attack the vortex burst moves forward and as it reaches the trailing edge of the wing lift is lost and a nose-up pitching moment is obtained. With increasing angle of attack, the vortex burst creeps forward on the wing, the lift curve slope decreases and drag rises steeply. Finally, at angles between 30° and 40°, maximum lift is obtained in combination with very high drag (fig. 2.23).

The nose-up pitching moment may become so high that even full-down elevon does not give sufficient nose-down moment for recovery. The delta wing is locked in a 'superstall' condition pitching up and down between 30° and 90° angles of attack as it 'pancakes' down through the air (fig. 2.24).

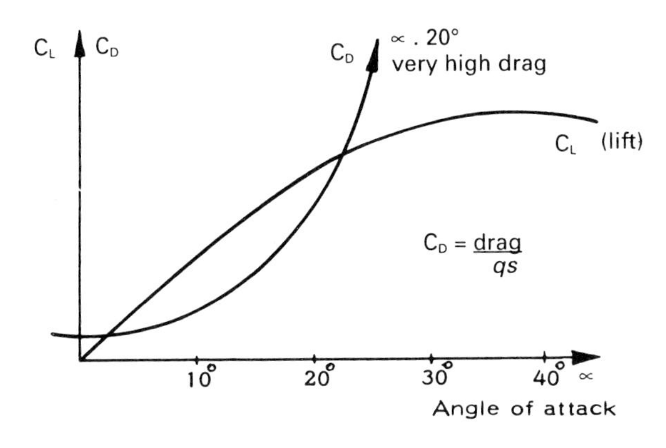

Fig. 2.23 *Delta wing lift and drag*

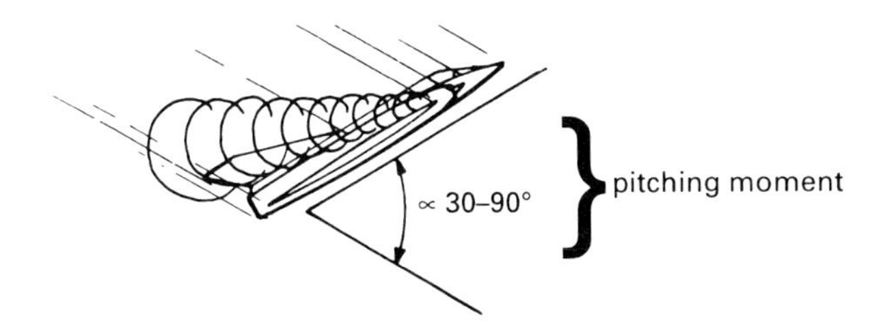

Fig. 2.24 *Delta wing locked in superstall*

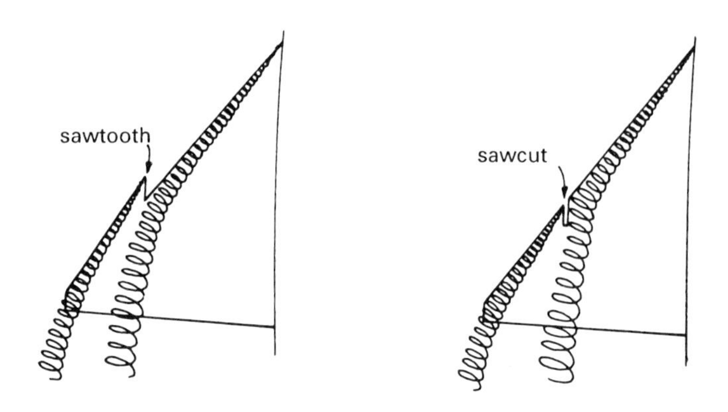

Fig. 2.25 *Leading edge discontinuities*

The vortex flow is sensitive to leading edge disturbances. This may be used to an advantage. If the wing is shaped as shown in fig. 2.25 the vortex from the wing-root drains off at the discontinuity where a fresh wing-tip vortex starts. This delays the vortex burst at the outboard part of the wing and increases the angle of attack where pitch-up is obtained. For wings in the medium-sweep range it is possible to use a combination of leading edge discontinuity and boundary layer fence to delay unpleasant flow disturbances.

Double-delta wings, such as the one in fig. 2.26, have a special problem. The strong vortex from the highly swept inboard wing section creates an upwash on the weaker outboard wing vortex. At intermediate angles of attack (15°–16°) the two vortices twist together, lift is lost on the outboard wing section and the aircraft pitches up. The vortex interference and the pitch-up can be delayed by cutting the outboard vortex into several vortices by means of the vortex dividers shown. The interference is now limited to one of the outboard wing vortices.

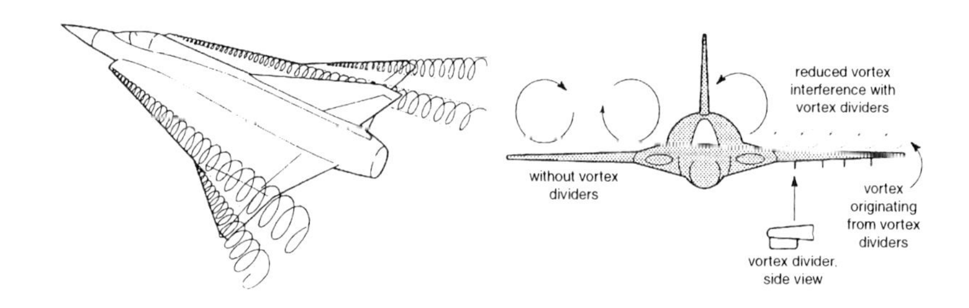

Fig. 2.26 *Double-delta without and with vortex dividers*

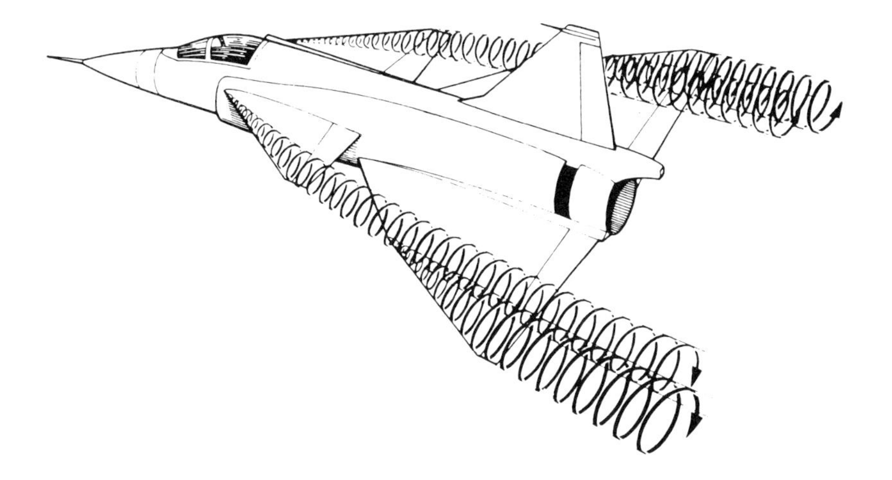

Fig. 2.27 *Vortex flow at canard-delta*

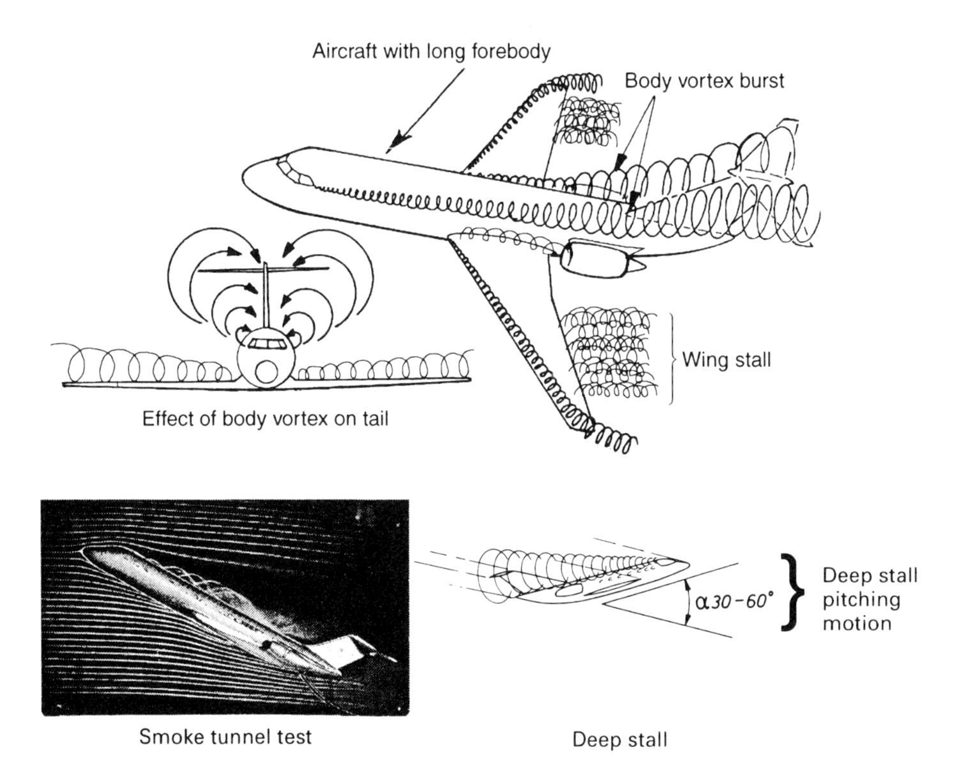

Fig. 2.28 *Body vortices and deep stall*

The vortex flows of the canard-delta is shown in fig. 2.27. The canard wing vortices are stabilised by the low pressure above the main wing, and vortex burst is delayed up to very high angles of attack. The main wing vortex downwash keeps the canard vortex near the main wing surface and prevents it from interfering with the fin. Main wing vortex burst is obtained as for plain deltas but the pitch-up is gentle and locked-in 'deep stall' can be prevented with sufficiently large elevons.

A long slim body may be regarded as a wing with an extremely high sweep angle. When the body reaches a certain angle of attack the flow separates into vortices even here. Aircraft with swept wings and long forebodies (typical for aircraft with body-mounted engines) may develop vortex flow both on the wings and above the fuselage, as shown in fig. 2.28.

Deep stall is caused by body vortices creating downwash on the horizontal tail of T-tailed aircraft. If this downwash becomes sufficiently large it is not possible to counteract the download on the tailplane with full-down elevator. The aircraft is locked in a deep stall. Other factors contributing to deep stall are wing-tip stall of the main wing and possibly some wake effects from the wing-root section on the engines. The body vortices usually interfere with the fin creating side forces and yawing motions. A 'Dutch roll' oscillation develops.

Local flow separations or flow disturbances may be obtained on swept wings at angles of attack well below stall. They give stability disturbances (especially in pitch) and increase drag. Typical cases of local flow disturbances are shown in fig. 2.29.

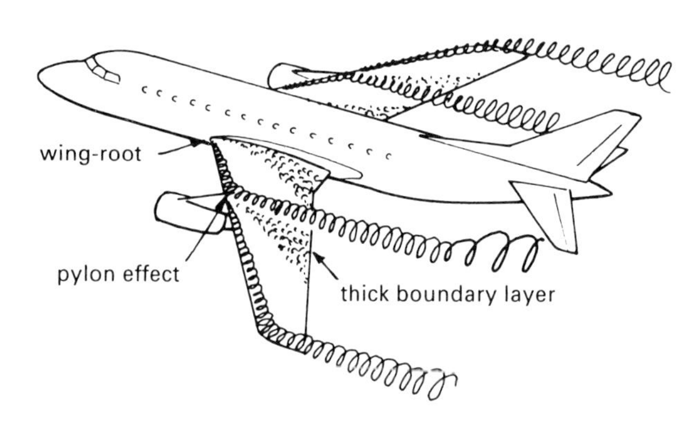

Fig. 2.29 *Local flow disturbances*

The wing-root section of an aircraft is always critical and may start separating early (poor design). The intersection between a wing and an engine pylon may affect the leading edge vortex flow which may result in a local separation. Furthermore, in the corner between the wing and pylon there is a pressure increase which tends to sweep the oncoming flow upwards and thus create a large local angle of attack resulting in a premature local flow separation. The dead air from the local separation may spread over the wing and thus induce a tip stall with pitch-up and drag increase.

For military aircraft, external stores protruding ahead of the wing may induce flow changes at intermediate angles of attack affecting the longitudinal stability and the drag of the aircraft. Even for clean wings it is possible for the flow condition to change fairly suddenly and then reattach at a higher angle of attack. This is especially the case for intermediate swept low aspect ratio aircraft with fairly unstable vortex flows.

It also happens that flow from the wing, such as a rearward-flowing vortex or a wing wake, affects the flow over the tail and induces pitching moment disturbances well below the wing stalling angle (fig. 2.30).

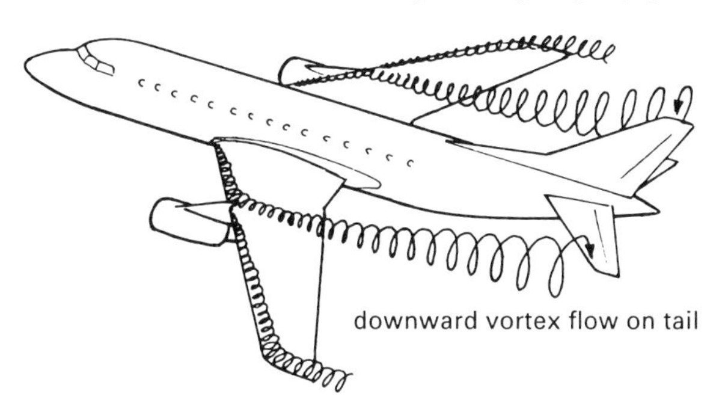

Fig. 2.30 *Wing vortex striking tailplane*

In other cases, it is possible for body wing-root vortices to strike the fin and eliminate directional stability through induced side sweep (fig. 2.31). For this reason two fins are often used.

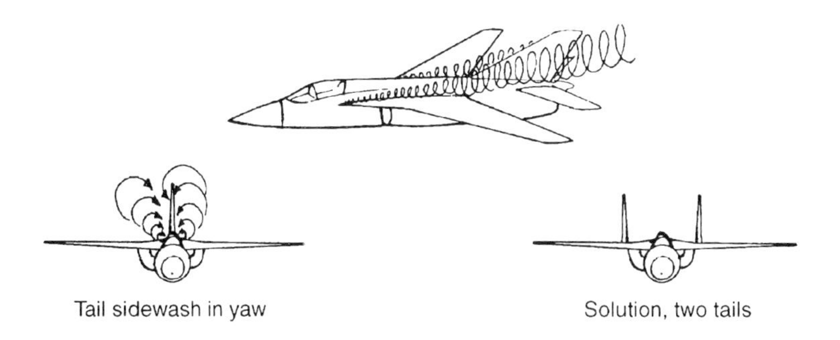

Fig. 2.31 *Wing vortex striking vertical fin*

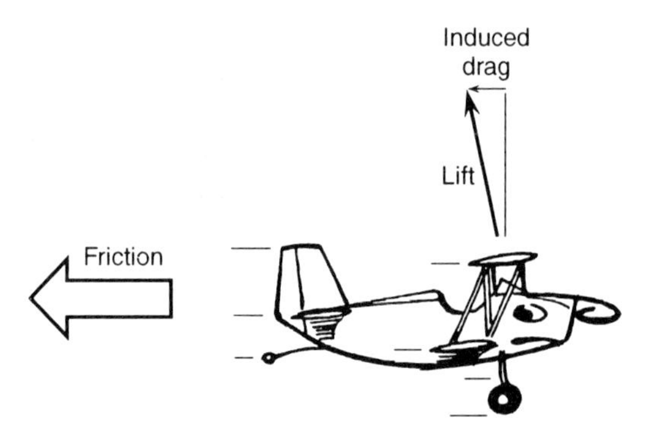

Fig. 2.32 *Aircraft drag components*

Drag

Two types of drag must be overcome in flight:

- drag due to friction
- induced drag due to lift

Friction Drag

Friction has two effects. Firstly, a friction force is obtained. Secondly, the flow separation of the boundary layer caused by friction creates a low-pressure wake behind the aircraft. The wake prevents the air from slowing to zero speed at the rear stagnation point. As a result, a rearward-acting suction force is obtained, instead of the forward-acting stagnation pressure obtained in frictionless flow (fig. 2.33).

The wake drag of blunt bodies (fig. 2.33b) is much higher than the friction drag. For this reason bodies with smooth surfaces and early laminar flow separation (giving a wide wake) have higher drag than blunt bodies with turbulent boundary layers and reduced wake thickness – even if the laminar flow friction force is much lower than turbulent layer friction. It is for this reason that golfballs have turbulence-inducing dimples (fig. 2.34).

Streamlined bodies like the wing profile in fig. 2.35a have mainly friction drag, providing the surface is maintained smooth and clean. Surface wear

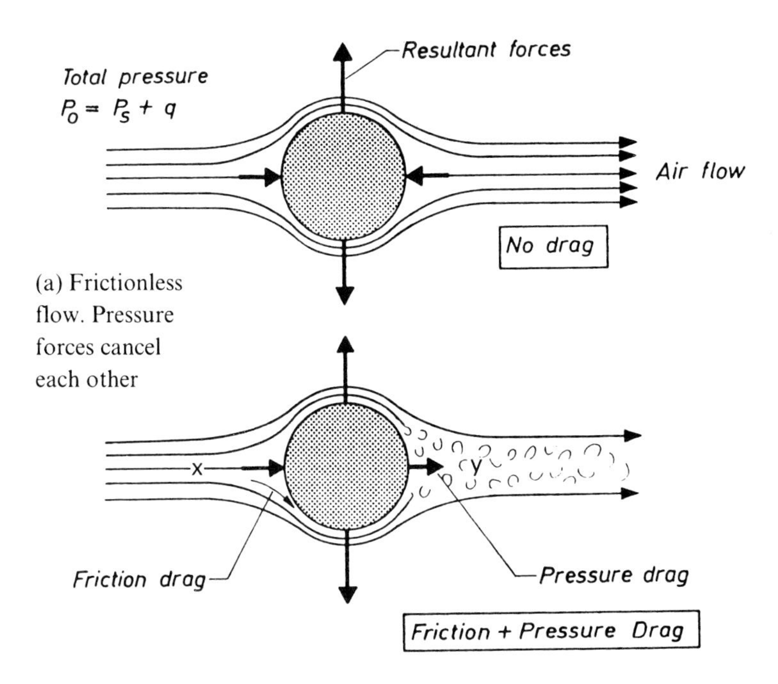

Fig. 2.33 *Friction drag of blunt body*

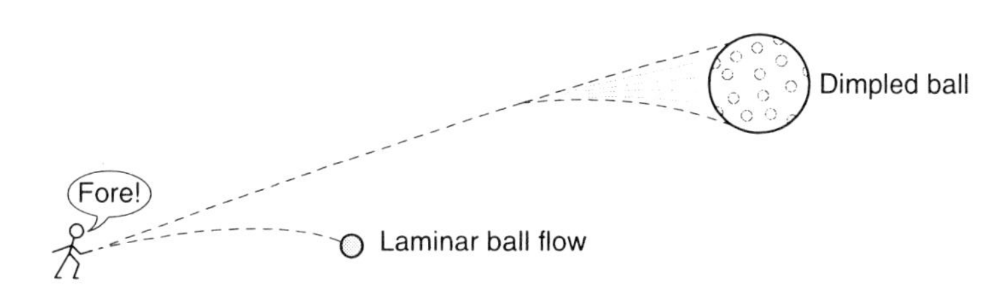

Fig. 2.34 *Dimpled golfball*

and contamination may, however, increase both friction and wake drag and more than double the original drag (fig. 2.35b). It is difficult to maintain sufficiently smooth aircraft surfaces in practical operation to prevent boundary layer turbulence.

However, special laminar flow profiles are made. These have their maximum thickness further aft on the chord than conventional profiles. As

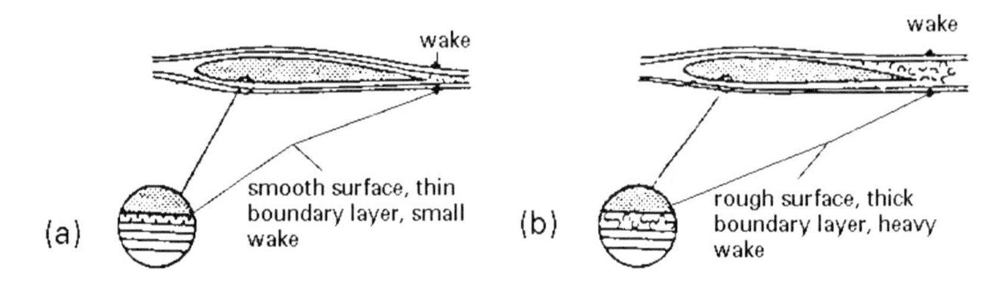

Fig. 2.35

a result, the minimum pressure point is moved rearward and a negative pressure gradient stabilising the laminar boundary layer flow is obtained, providing the surface is kept extremely clean. The problem with laminar flow profiles in practical operations is that surface contamination may trigger an abnormal growth of the boundary layer thickness behind the low-pressure point resulting in a thick, turbulent wake. The drag of the so-called low-drag laminar profile then becomes higher than the drag of a more conventional profile.

The friction drag of an aircraft is the sum of the friction forces plus the sum of flow separation drags at all surface discontinuities at air inlets and outlets, antennas, and steps in the surface between wings, flaps and control surfaces etc. Flaps and control surfaces that are installed out of tolerance, surface wear and contamination, and air leakage from pressure cabins into the airflow may have large effects on the drag of the aircraft (fig. 2.36 and fig. 2.37).

Friction drag depends on the speed, density and volume of the air flowing over the aircraft surfaces and, therefore, increases with the square of the speed (fig. 2.37).

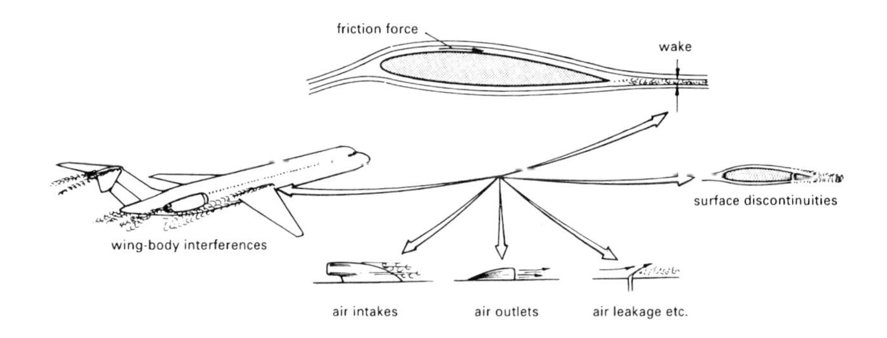

Fig. 2.36 *Aircraft drag due to friction*

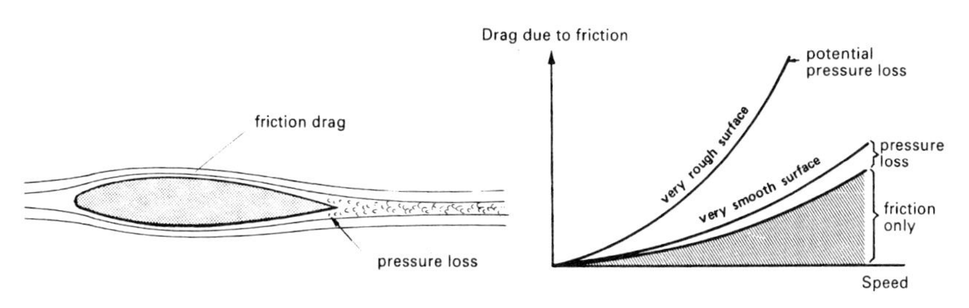

Fig. 2.37 *Effect of speed on friction drag*

Induced drag

Infinite-span wings have constant lift along the span and no induced drag since the lift resultant is perpendicular to the direction of flight (i.e. to the approaching undisturbed flow). Thus, lift equals the normal force N, the resultant of the pressure distribution normal to the local flow (fig. 2.38). Wings with finite spans lose lift at the tips as air flows from the high-pressure region below the wings to the low-pressure region above and form tip vortices (fig. 2.38b).

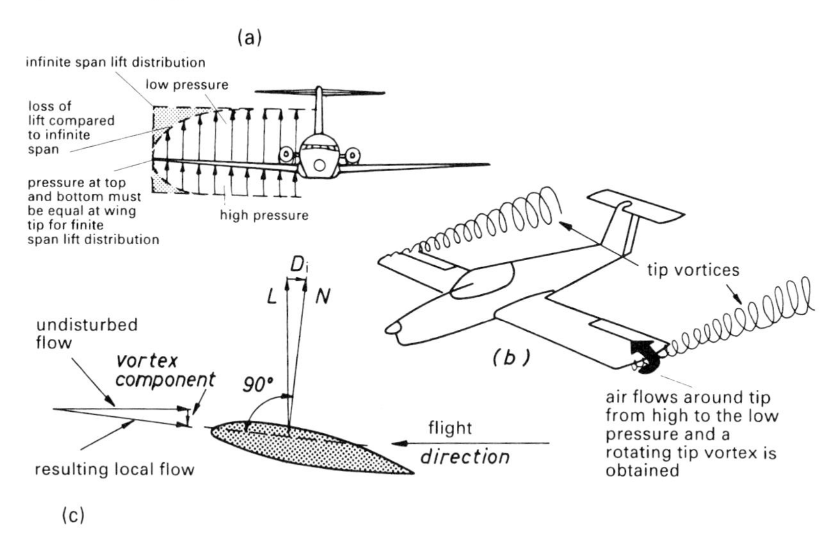

Fig. 2.38 *Induced drag*

The tip vortices induce a downward flow component along the span reducing the local angle of attack (fig. 2.38c). This tilts the normal force N rearward, reducing lift L and creating an induced drag component D_i.

Since the effects of the vortex are greatest near the wing-tips, the induced drag increases when the wing aspect ratio decreases. As the aspect ratio is reduced, the rearward tilt of the normal force increases. Thus, at a given angle of attack, lift decreases and induced drag increases. Theory proves that induced drag is inversely proportional to the aspect ratio. A delta wing may, therefore, have induced drag which is six times as high as that of a soaring aircraft wing.

Wings with elliptical planforms have the lowest induced drag. As the relative length of the tip chord increases, induced drag increases somewhat. A rectangular wing with an aspect ratio of 6 has 5% higher induced drag than an elliptical wing. Wings with taper ratios around 0.5 have only 1% higher induced drag than elliptical wings.

Since increased lift requires increased angle of attack,which tilts the normal force further aft, induced drag becomes proportional to lift squared (fig. 2.39).

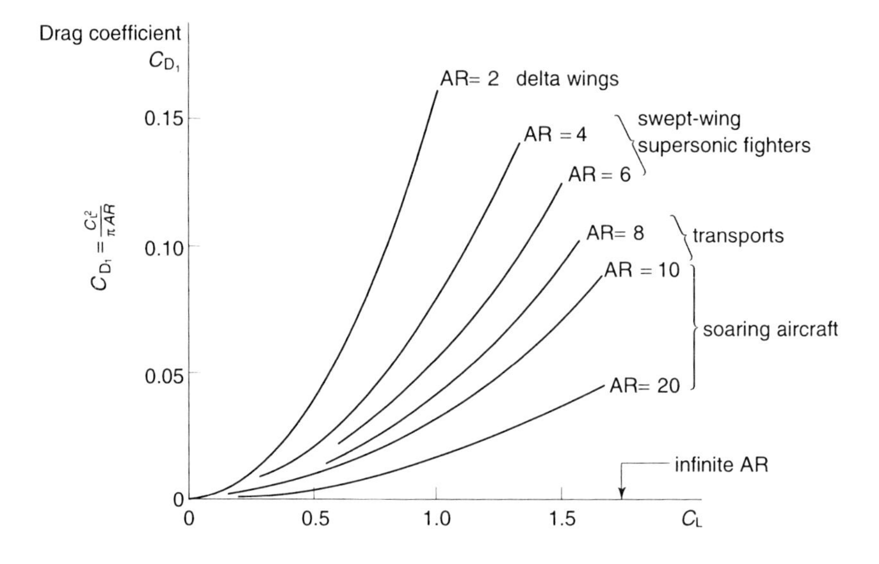

Fig. 2.39 *Effect of lift and aspect ratio on induced drag*

When the aircraft's speed is reduced and lift is kept constant (lift, L is equal to the aircraft's weight, W) the angle of attack of the wing increases twice as fast as speed is reduced. Hence the induced drag is quadrupled when flight speed is halved. This can be written:

induced drag $D_i = C_{D_i} qs$

$$= \frac{C_L^2}{\pi AR} qs$$

but since $C_L = \dfrac{L}{qs}$

$$D_i = \frac{L^2}{\pi\, ARq^2s^2}\, qs$$
$$= \frac{n^2_z\, W^2}{\pi AR\, qs}$$
$$= \frac{Kn^2_z W^2}{\pi V^2}$$

where K is a constant depending on aspect ratio, wing area and air density.

For a given aircraft this means that the induced drag increases with the square of speed reduction and directly with the square of aircraft's weight W and load factor n_z. Increasing weight by 20%, increases induced drag by 44%. Pulling two g quadruples the induced drag (fig. 2.40).

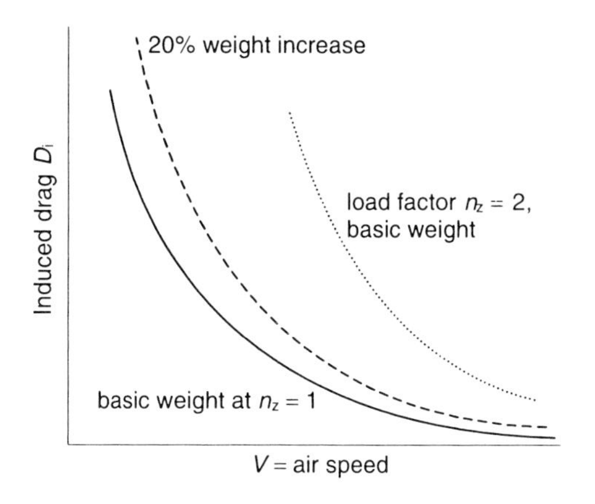

Fig. 2.40 *Effects of speed, weight and load factor on induced drag*

Total Drag

Combining induced drag and friction drag, the total drag in fig. 2.41 is obtained. The figure shows that aircraft drag increases towards infinity at low speed due to high induced drag, decreases to a minimum at an intermediate speed, and then increases again due to increasing friction drag.

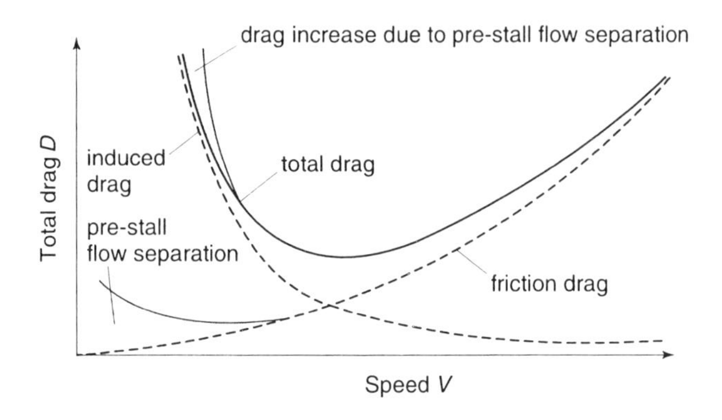

Fig. 2.41 *Drag versus speed*

Ground Effect

When an aircraft flies very close to the ground the pressure between the wing and the ground increases. This increases the upwash in front of the wing. The effect is the same as if the downwash from the tip vortices decreased because it cannot flow through the ground. The resultant increase of the local angle of attack increases the leading edge suction and tilts the normal force component forward decreasing the induced drag (fig. 2.42). Slight drag reductions due to ground effect are obtained approximately one wing span above the surface. Pronounced effects are obtained up to a height of twice the landing-gear height above the ground.

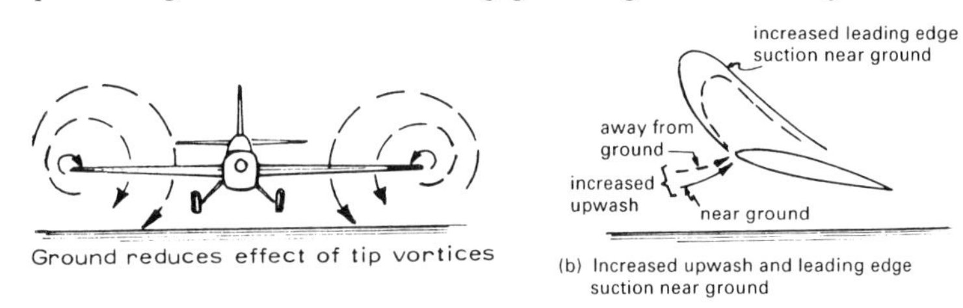

Fig. 2.42 *Ground effect on induced drag*

High-lift Devices

Movement of control surfaces, flaps and spoilers changes the pressure distribution around the whole wing profile (fig. 2.43). Flap operation increases the pressure below the wing. This moves the forward stagnation point aft and increases the upwash in front of the wing. As a result, the leading edge suction and the upper-surface lift both increase. At the flap,

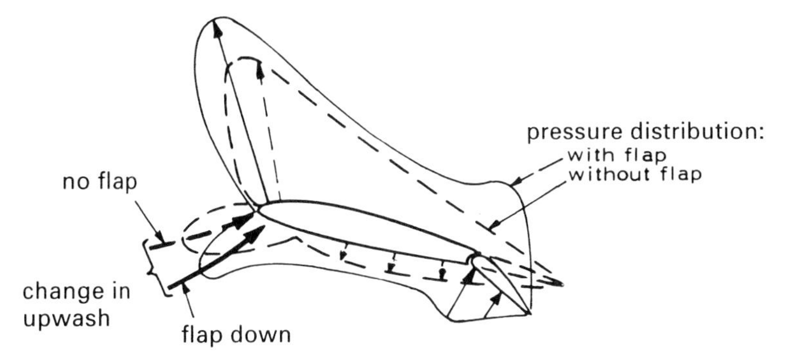

Fig. 2.43 Effect of flap angle on pressure distribution

suction increases when the air flow is accelerated around the knee (fig. 2.43).

The flap increases the downward deflection of the airflow at constant angle of attack (fig. 2.44).

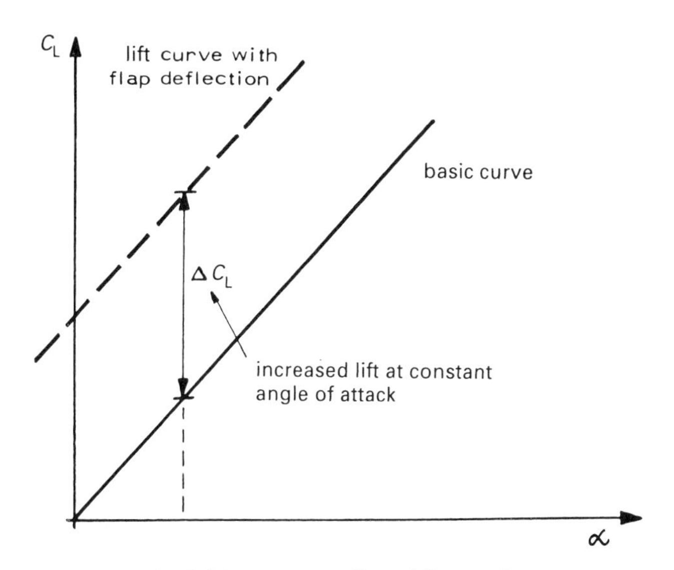

Fig. 2.44 *Primary effect of flap angle*

The magnitude of ΔC_L, flap efficiency, is a function of

- flap chord
- flap span
- flap efficiency and extension angle.

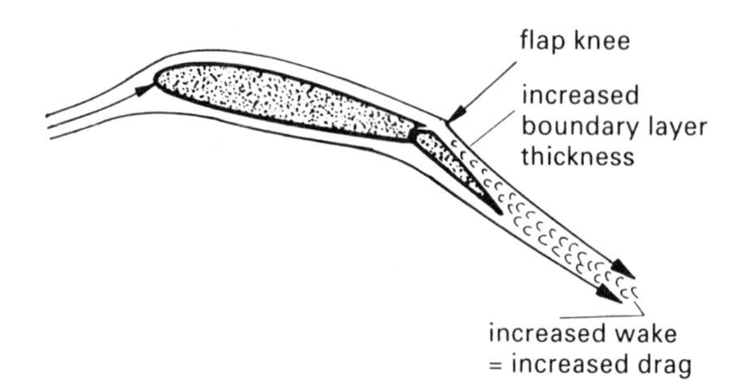

Fig. 2.45 *Boundary layer thickness increase at flap knee*

Flap efficiency is a function of the boundary layer behaviour as it flows past the flap knee (fig. 2.45).

The boundary layer thickness increases sharply as the flow runs through the increasing pressure past the flap knee. As the flap angle increases so does the boundary layer and gradually a flow separation is obtained that makes further flap movement fruitless. The increasing wake thickness reduces flap efficiency and increases aircraft drag. Typical curves of ΔC_L versus flap deflection and ΔC_D versus flap angle are shown in fig. 2.46.

The efficiency of the flap, and thus the flap angle at which maximum ΔC_L is obtained, can be improved by re-energising the boundary layer by:

- the use of one or more slots where high-energy air from the lower surface can flow into and mix with the upper-surface boundary layer
- blowing high-pressure engine air into the boundary layer
- sucking away part or all of the boundary layer.

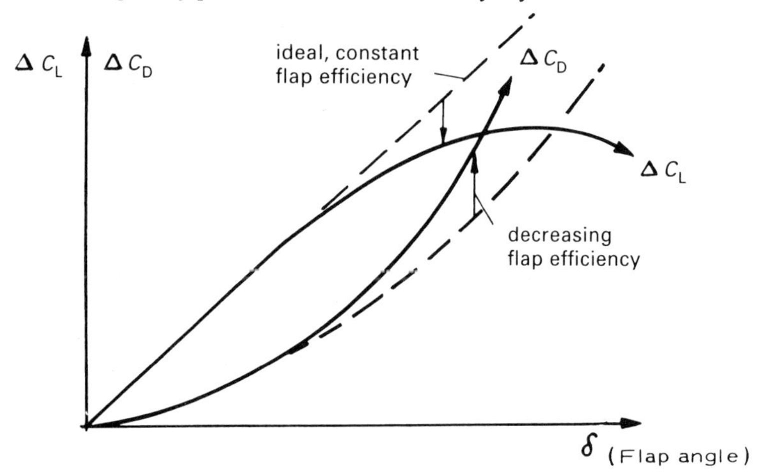

Fig. 2.46 *Lift and drag increase with flap angle*

This does not change the general picture that flap efficiency decreases and drag increases with increasing flap angle.

Note that there may be some risk for sensitivity to small disturbances for highly efficient flap systems, which means that small disturbances may result in large losses of lift and big increases in drag.

The ΔC_L, the flap efficiency, remains fairly constant with angle of attack up to the angle where the pressure distribution over the profile nose is fairly equal to the one that resulted in stall of the plain profile, as shown in fig. 2.47. At this angle, the boundary layer of the plain profile is very close to separation and the boundary layer reaching the flap (fig. 2.47b) is in a very critical condition. Now, since the flap angle increases the local angle

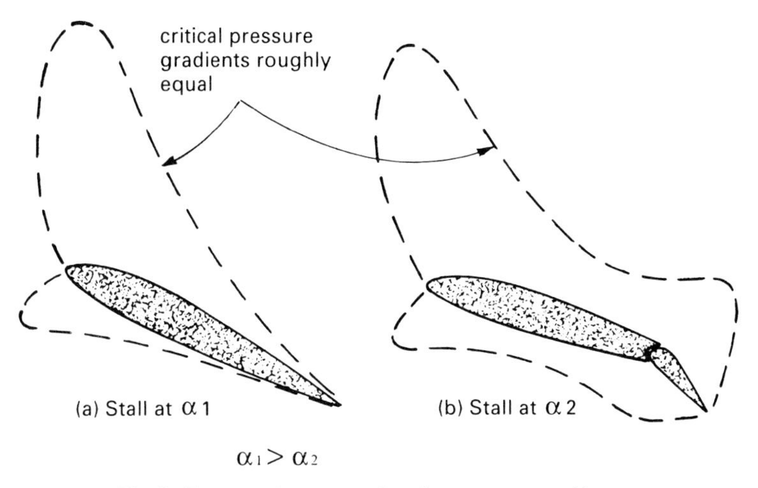

Fig. 2.47 *Equal pressure distribution over profile nose*

of attack at the profile nose, the angle of equal pressure distribution over the nose (α_2 in fig. 2.48) decreases with increasing ΔC_L (i.e. with increasing flap chord, flap angle and flap efficiency). Therefore, the stall angle decreases with trailing edge flap deflection (fig. 2.48).

Increased stall angle can be obtained by changing the pressure distribution over the profile nose and by re-energising the boundary layer, for instance, by means of nose flaps or slats (fig. 2.49). A drooped profile nose decreases the suction peak and thus the pressure gradient (pressure increase) behind the nose. This delays boundary layer separation. Further delay is obtained if the droop is combined with a slot permitting fresh air from the lower surface to mix with and re-energise the upper surface boundary layer.

The immediate effect of lowering the wing nose at low angles of attack is, however, to reduce lift and increase drag. At very low angles of attack,

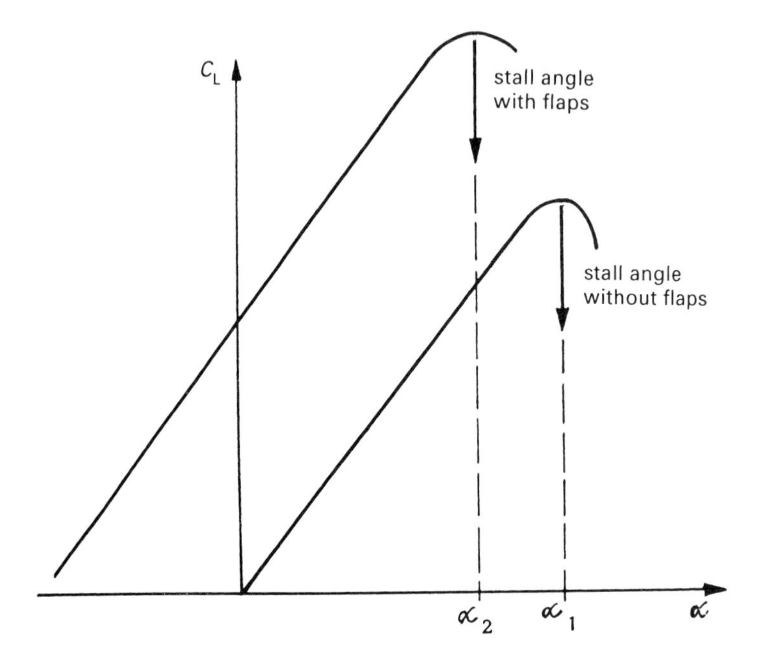

Fig. 2.48 *Effects of flaps on stall angle*

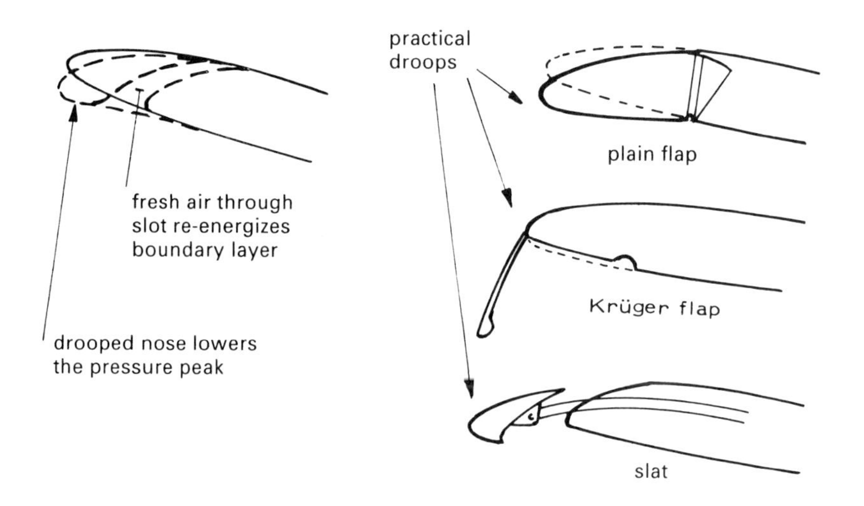

Fig. 2.49 *Profile nose devices*

flow separation is obtained at the lower side of the wing nose with decreased lift and increased drag as a result (fig. 2.50).

A spoiler has the opposite effect of a flap. If it is extended when the flaps are down, the combination gives a very large drag increase (fig. 2.51). The flap increases the lift for landing, the spoiler cancels the lift, and the combination gives high drag.

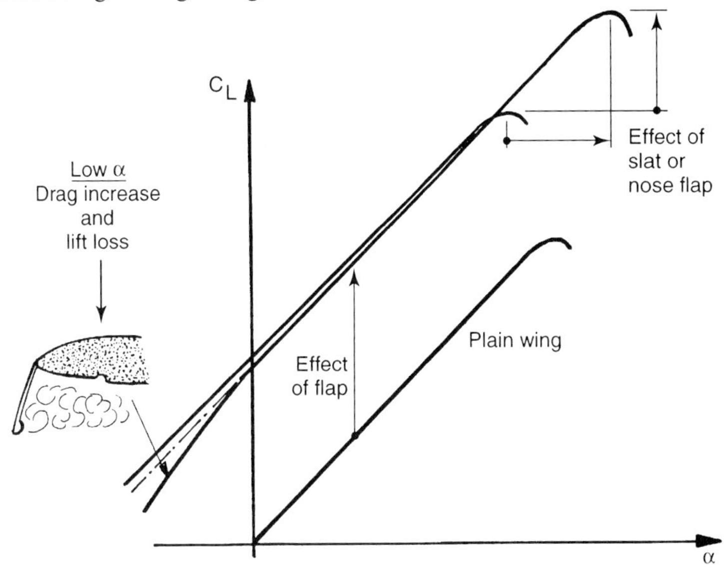

Fig. 2.50 *Effect of leading edge devices on lift*

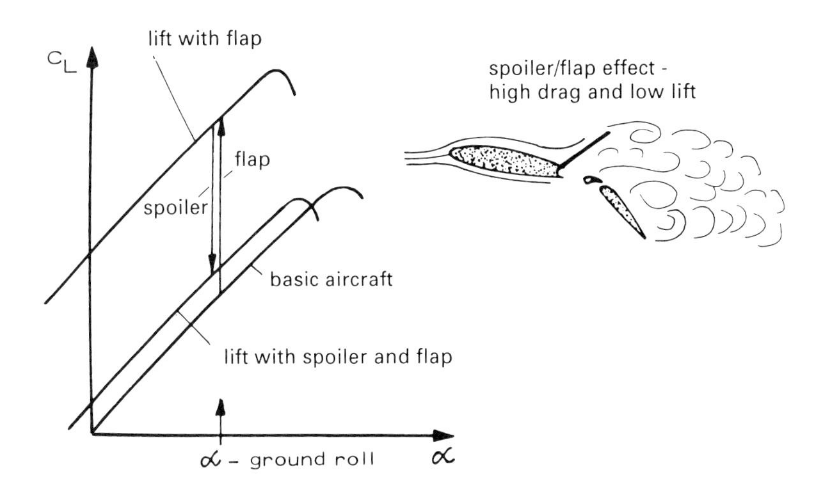

Fig. 2.51 *Spoiler effect on lift*

The spoiler may also be used to some extent to increase drag for rapid descent. However, the angles used are much lower than the lift-dumper angles used for landing. Spoiler operation increases the stalling speed. For maximum angles the increase is often very large.

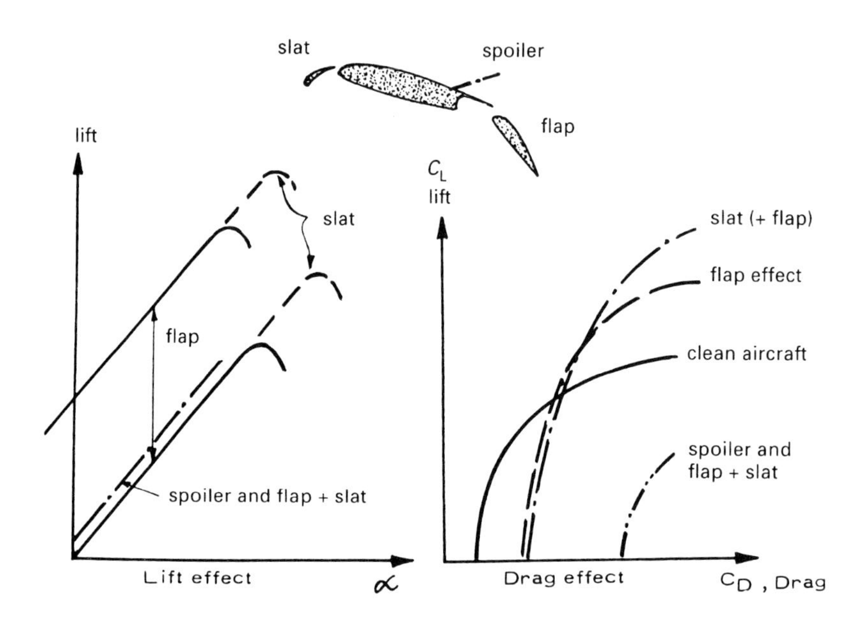

Fig. 2.52 *Typical aircraft lift and drag data*

Let us now look at a typical set of aircraft lift and drag data including the effects of high-lift devices and spoilers (fig. 2.52).

High-lift devices increase lift and drag. However, the drag of an aircraft with high-lift devices extended may be much less than the drag of the clean aircraft near to stalling and definitely less than for the stalled clean aircraft.

Spoilers increase drag, decrease lift and increase the stalling speed.

Recalculated into thrust required for flight, the drag data in fig. 2.52 look as shown in fig. 2.53.

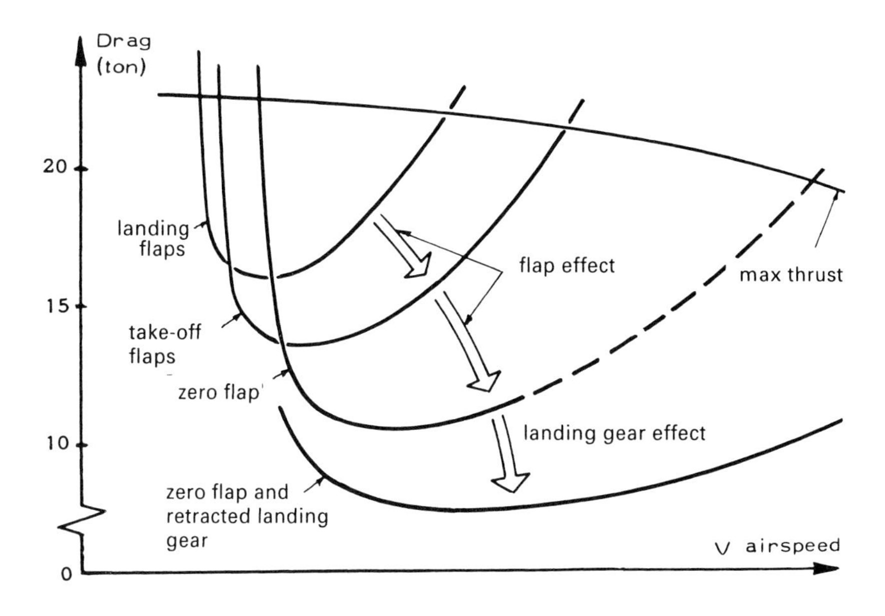

Fig. 2.53 *Drag versus speed for various aircraft configurations*

CHAPTER 3

TRANSONIC AND SUPERSONIC AERODYNAMICS

Supersonic Pressure Distribution

There is a fundamental difference between subsonic and supersonic flow. Since pressure disturbances travel with the speed of sound, it is possible for the flow at subsonic speed to react to the pressure changes around the body before reaching them (fig. 3.1). The flow in front of a wing sweeps up away from the high stagnation pressure towards the low pressure above the wing.

At subsonic speeds, therefore, a peaky pressure distribution is obtained with the resultant lift force along a line located roughly at 25% of the profile chord.

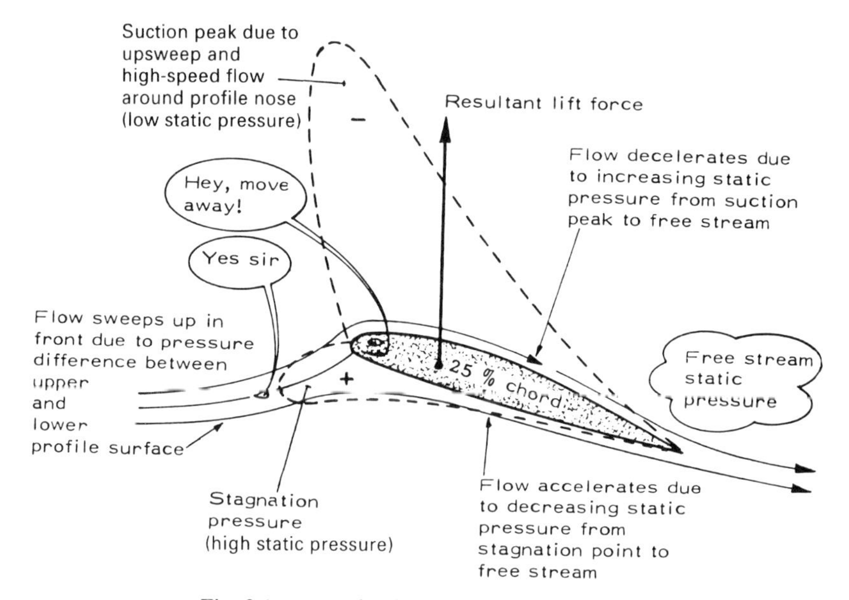

Fig. 3.1 *Interplay between flow and pressures*

At supersonic speed, however, the body travels faster than the pressure disturbances and all flow changes take place suddenly at the leading or trailing edges of the body, or at places where the body changes curvature (fig. 3.2 and fig. 3.3). There are no changes in the flow along flat surfaces since the flow feels no pressure changes in the flow direction (fig. 3.3).

At supersonic speeds, pressure reductions occur through expansion

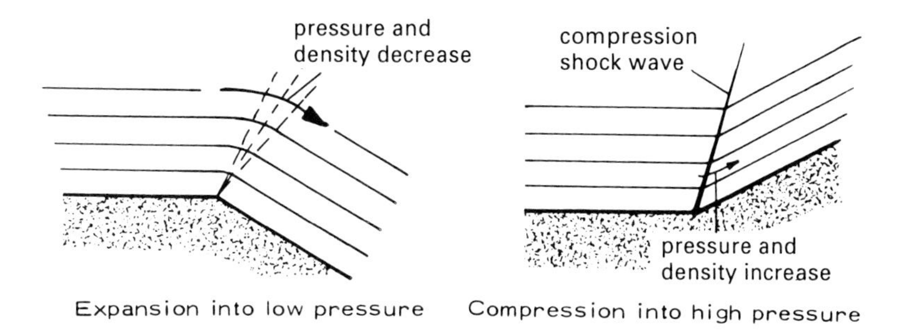

Fig. 3.2 *Supersonic expansion and compression*

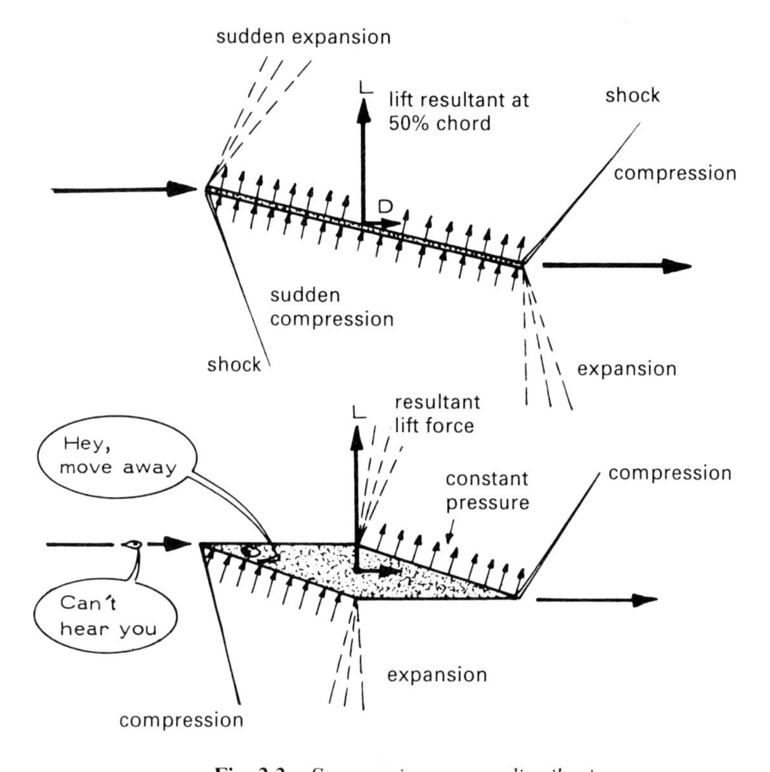

Fig. 3.3 *Supersonic pressure distribution*

waves. Pressure increases are, as a rule, sudden and occur through one or more shock waves. The pressure distribution changes from the peaky subsonic type to rectangular shape and the resultant lift force moves from the 25% line on the chord to the 50% line.

Transonic Flow Disturbances

As an aircraft approaches the speed of sound, local sonic velocities may be reached at several places. This may have several effects (fig. 3.4a). Fig. 3.4b shows shock wave formation in air around a Phantom flying at transonic speed. The humid air condenses as the air reaches local supersonic speed and the condensation disappears suddenly as the air is compressed to subsonic speed through the shock wave.

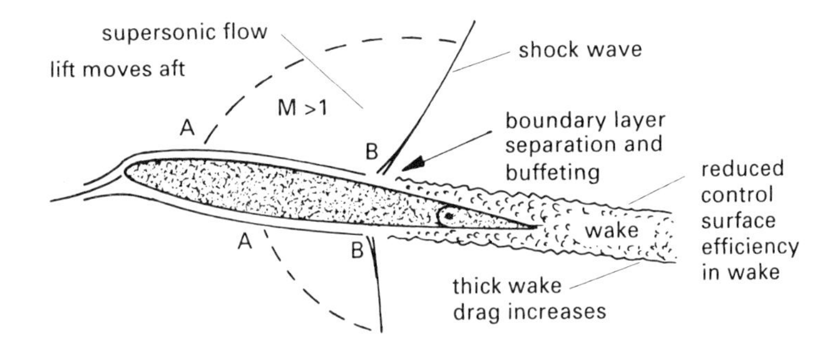

Fig. 3.4a *Shock-induced boundary layer separation*

Fig. 3.4b

Supersonic flow may be reached at a point A in fig.3.4a. The flow continues to accelerate as it expands round the curvature of the profile. A sudden decrease to subsonic speed occurs through a shock wave at some point B. As the speed increases, point A moves forward and the shock wave moves aft towards the trailing edge.

Shock waves cause boundary layer separation due to the large pressure increase through the wave. This has the following effects:

- drag increases due to increased wake thickness (increased pressure drag)
- turbulence in the wake may result in buffeting of the aircraft and control surface buzz
- the efficiency of control surfaces (ailerons, elevators and rudders) located in the wake may be sharply reduced
- shock waves may oscillate and induce oscillations in the control surfaces and wings
- the wing stall angle (maximum lift) may be sharply reduced.

At high angles of attack, local supersonic flow may be obtained at the wing leading edges at low Mach numbers. The loss of maximum lift due to shock-induced boundary layer separation, therefore, begins at roughly twice the take-off and landing speeds. The maximum lift change of a swept-wing aircraft as a function of Mach number shown in fig. 3.5 is typical.

For the aircraft in question, maximum lift is reduced by 80%.

When the local supersonic flow at high Mach numbers approaches the

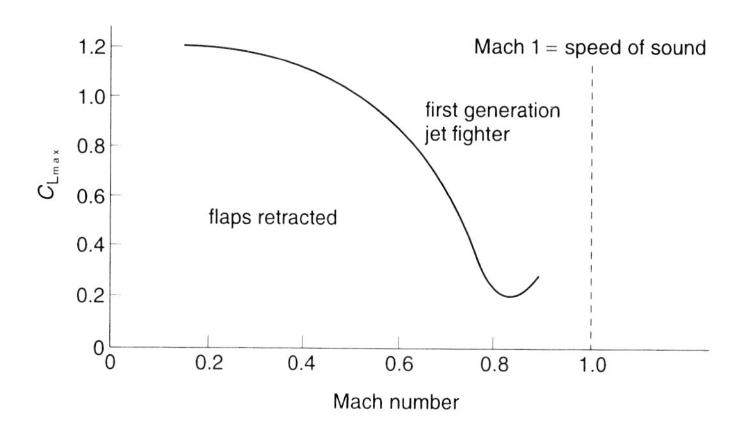

Fig. 3.5 *Maximum lift versus Mach number*

wing trailing edge, the shock-induced flow separation occurs far back on the profile and the effect on stall decreases. Lift therefore begins to increase again.

Rearward Shift of Wing Lift

Rearward movement of the lift resultant due to a change from subsonic to supersonic pressure distribution was shown in fig. 3.1 and fig. 3.3. The shift of the centre of pressure begins in the transonic region as soon as local supersonic flow is obtained.

The result is an increase in aircraft stability as speed increases from subsonic to supersonic (fig. 3.6). Consequently, the trim force required for pitching moment trim and for manoeuvres increases at supersonic speeds.

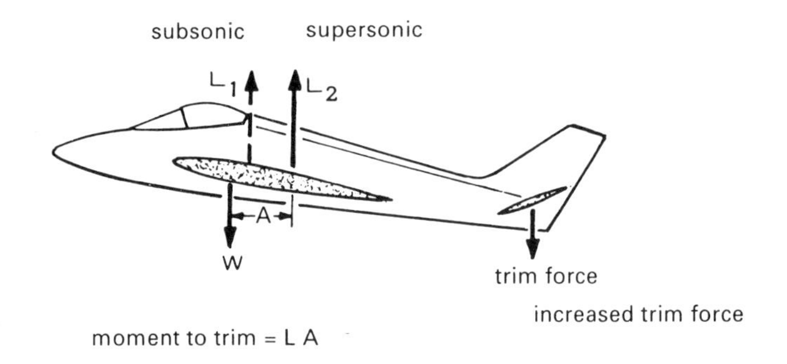

Fig. 3.6 *Increased stability at supersonic speed*

Reduced Control Surface Efficiency

Control surface efficiency decreases in supersonic flow (fig. 3.7). At subsonic speeds the operation of an elevator changes the lift of the whole tail. At supersonic speeds, however, the flow ahead of the elevator does not feel the pressure changes due to elevator movement and therefore the effect of an elevator movement is much smaller at supersonic than at subsonic speeds.

Another effect of supersonic speed is that the centre of pressure on the control surface moves rearward due to the change from triangular to rectangular pressure distribution. Thus, the moment (and hence the stick force) required to move the elevator increases at supersonic speed.

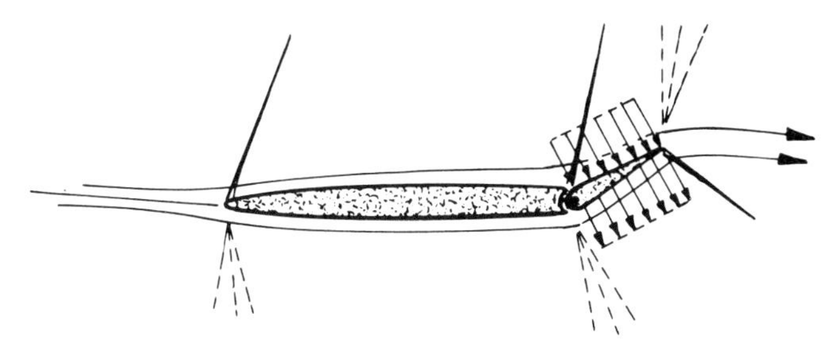

Fig. 3.7 *Pressure change due to elevator angle*

Thus at supersonic speeds:

- increased stability requires increased trim force
- reduced control surface effectiveness requires a large increase in control surface angle to obtain the increased trim force
- rearward movement of the resultant force on the control surface increases the moment required to rotate the surface.

Thus, the stick force required to trim or manoeuvre an aircraft at supersonic speed may grow beyond control.

For modern supersonic aircraft, these problems are reduced by using powerful hydraulic control systems and all-flying tails or by designing aircraft with negative subsonic stability and artificial stabilisation systems. This does not mean that it has been possible to avoid reductions in manoeuvrability from subsonic to supersonic speeds.

Transonic and Supersonic Drag

Special drag effects are obtained at supersonic speeds even in frictionless flow (which gives zero drag at zero lift at subsonic flow) due to the continued expansion (and hence pressure reduction) of the air as it flows towards the trailing edge of the profile (fig. 3.8).

This drag increase begins as soon as supersonic flow is reached on the body and combines at transonic speeds with the drag increase due to shock-induced boundary layer separation. As a result, there is a sharp increase in drag in the transonic region until the shock waves have reached the trailing edge of the body and the zero-lift increase in drag again becomes approximately proportional to the dynamic pressure, as it is at subsonic speed (fig. 3.9).

Wing sweep is one method used to delay transonic flow problems. This works as shown in fig. 3.10. The velocity component in the flight direction

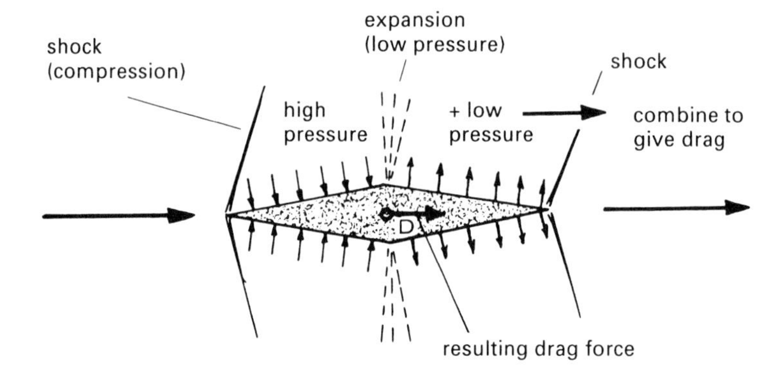

Fig. 3.8 *Supersonic wave drag*

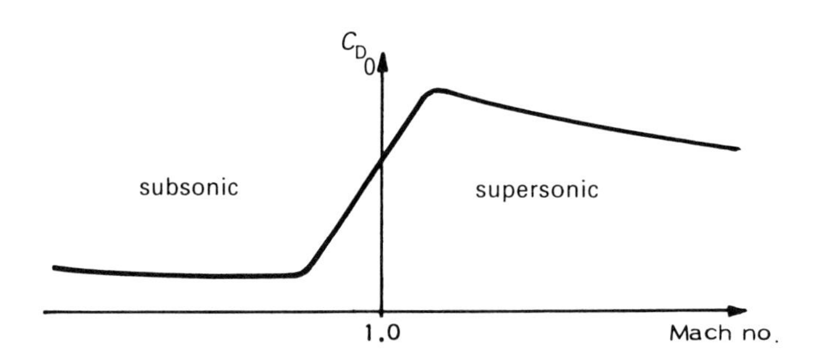

Fig. 3.9 *Zero lift drag coefficient*

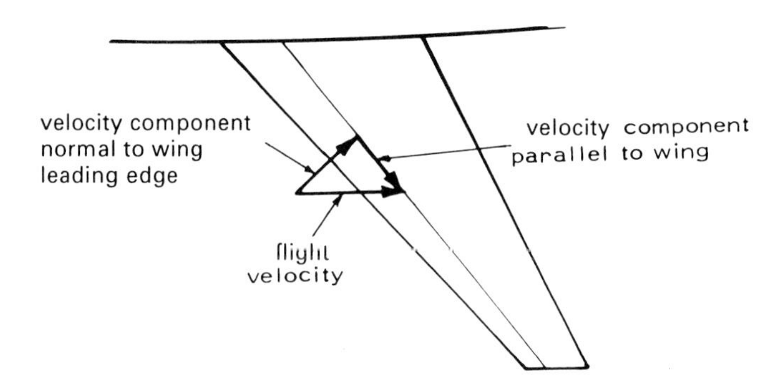

Fig. 3.10 *Effect of wing sweep*

may be considered to consist of two components, one normal to the wing leading edge and one parallel to the leading edge. Since the flow in the span-wise direction does not change speed, the drag increase comes from the component normal to the leading edge, and this is much less than the flight speed. The change in drag with sweep is illustrated in fig. 3.11.

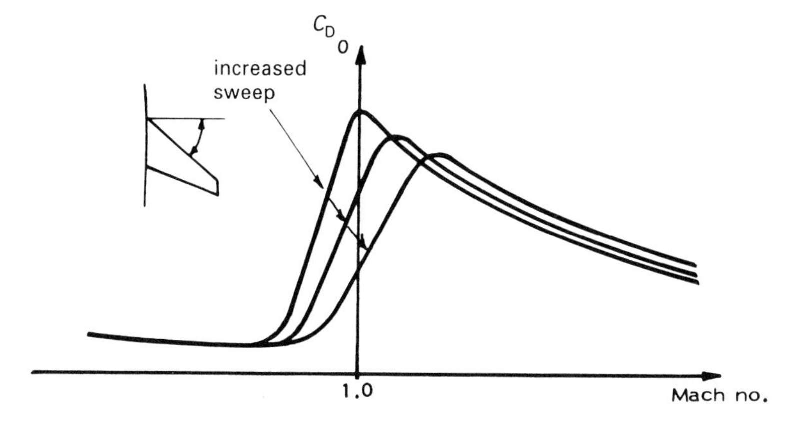

Fig. 3.11 *Effect of wing sweep on wave drag*

Another drag component that must be considered at supersonic speeds is the drag due to trim. Large elevator, tailplane or elevon angles are required to trim the large nose-down moment caused by the rearward shift of the lift resultant and the reduced control surface efficiency at supersonic speeds (fig. 3.12).

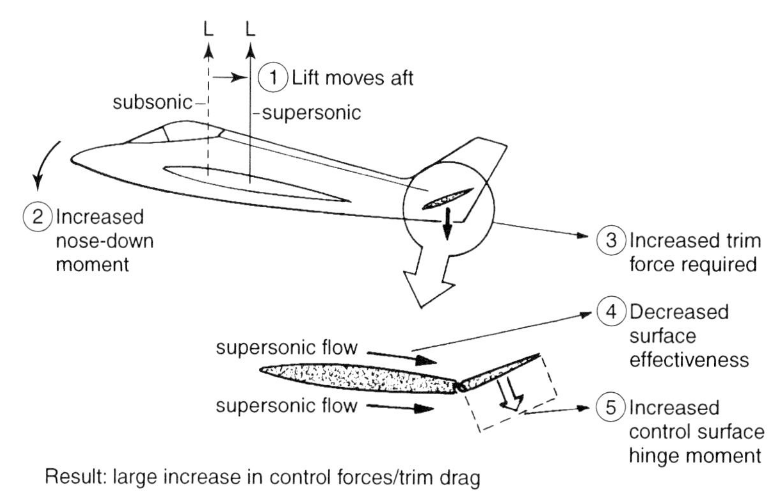

Fig. 3.12 *Control-reducing effects add up at supersonic speeds*

The supersonic trim drag may be very large. For supersonic transports (Concorde) and bombers it can be reduced by pumping fuel aft during supersonic cruise thus reducing the longitudinal stability.

Supersonic fighters and attack aircraft may be designed with negative longitudinal stability at subsonic speeds in order to decrease supersonic trim drag and increase manoeuvrability. Advanced, electronic stability and control systems are then required to obtain acceptable subsonic aircraft handling qualities.

Flight into the supersonic region increases aircraft stability and drag, reduces aircraft manoeuvrability and may result in a series of transonic disturbances due to shock-induced flow separations.

CHAPTER 4

HANDLING QUALITIES

Introduction

The ability of a man to control a motion is affected by the complexity of the action, by his training, and by disturbing factors.

For example:

most men can throw one ball in the air and catch it again,

quite a few can manage to juggle several balls at the time

and

some can train themselves to do the most fantastic balancing and juggling acts

the simple control task may not be affected by an outside disturbance

but

the complex one may collapse completely. There is a limit to man's simultaneous control capacity

Fig. 4.1 *Simultaneous control capacity*

The ability of a man to control a movement is affected by the complexity of the movement, his training, and disturbance factors.

For example, most men can throw one ball in the air and catch it again, quite a few can juggle several balls and some can train themselves to do the most fantastic balancing and juggling acts. The simple control task may not be affected by an outside disturbance but the complex one may collapse completely. There is a limit to man's capacity for simultaneous control.

Fig. 4.2 *Simple control task*

Similarly, in a low-speed aircraft equipped for 'sunshine flights' only, the pilot workload is low. The aircraft deviates slowly from its trimmed condition and control of flight path or speed disturbances becomes easy – providing the control response is good. As a matter of fact, some of the well-known first aircraft were longitudinally unstable, but still 'acceptable' as flying machines.

Fig. 4.3 *Complex task with risk for overload*

In a high-speed aircraft equipped for instrument flight, the pilot workload may approach saturation while landing in severe weather. At the same time, the aircraft may deviate rapidly from trimmed flight due to small disturbances. In order to avoid control collapse the workload must be reduced. This can be done by introducing stringent stability and control requirements for aircraft design. Poor handling qualities can make it impossible for pilots to maintain control in critical situations.

In the following section, some of the stability and control requirements will be discussed.

Handling Qualities Requirements

The basic handling qualities requirements can be summarised as follows:

- it must be possible to trim an aircraft to zero stick force at all speeds providing the aircraft is correctly loaded
- the aircraft must return to trim after a disturbance
- the aircraft oscillations, following a disturbance, must be well damped
- the aircraft's response to control inputs must be quick enough for safe handling in turbulent conditions
- excessive control sensitivity must be avoided in order to prevent pilot-induced oscillations
- the control work, i.e. the combination of control forces and movements, must not cause undue strain and must be well balanced in all directions. Control end positions must be within easy reach
- the aircraft must be fully controllable down to stall and up to maximum dive speed and the control system must permit rapid recovery from stall and high-speed dives
- easily recognisable warnings are required for dangerous low-speed and high-speed flight conditions.

Today, good handling qualities requirements are available for most aircraft. For military aircraft and transports the preliminary designs are usually flown in advanced simulators and adjusted when necessary before they reach the full flight-test stage.

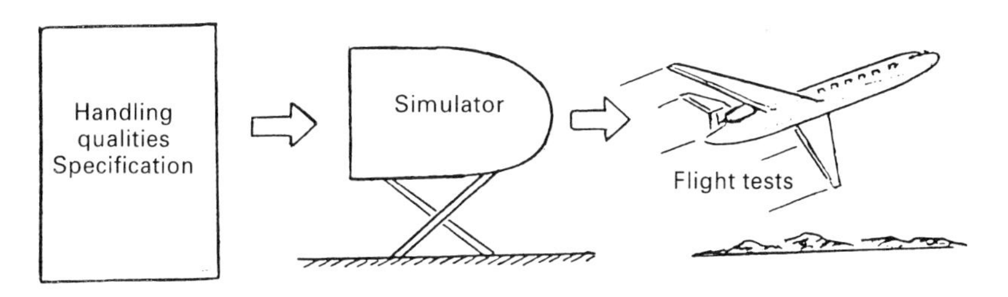

Fig. 4.4 *Handling-quality design*

However, all problems are not solved by satisfying these requirements. Accidents caused by loss of stability and control still happen.

Aircraft Motions

An aircraft in trimmed, level flight at constant speed can be quite 'restless' in all but absolutely still air. Turbulence induces oscillations and makes the aircraft move up, down and sideways about the trimmed flight path.

The discussions of the aircraft motions are simplified by dividing them into longitudinal, directional and lateral according to the axes illustrated in fig. 4.5.

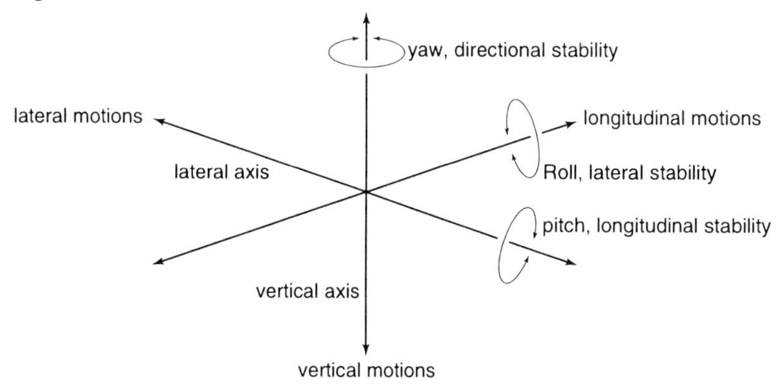

Fig. 4.5 *Stability and control axes*

With this coordinate system the aircraft's stability can be divided into three types:

- trim
- static stability, giving basic requirements for return to trim after a disturbance
- dynamic stability, describing the oscillations of the aircraft after a disturbance.

Longitudinal Trim, Stability and Control

Positive Longitudinal Stability

When an aircraft in flight is disturbed by a gust, it rotates about its centre of gravity (c.g.). The aircraft is stable when the lift resultant is located

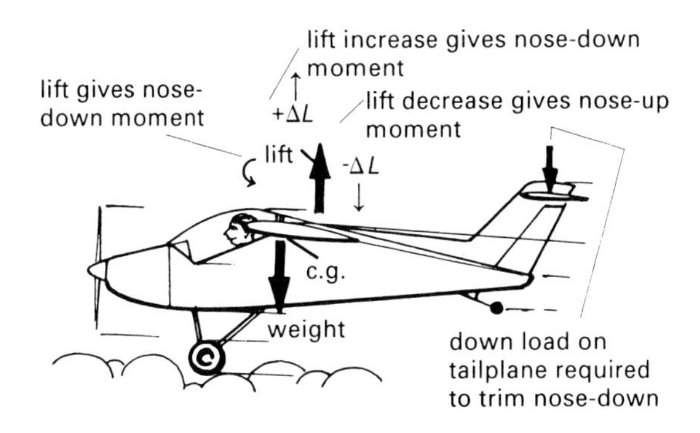

Fig. 4.6 *Positive stability with c.g. forward of lift resultant*

behind the c.g. and lift changes give moment changes returning the aircraft to trim after a disturbance (fig. 4.6).

A stable aircraft, returning to trim after a disturbance, reduces pilot workload and is pleasant to fly.

Control Movement

With the lift resultant located behind the c.g., a positive control movement is obtained for speed changes and manoeuvres. Pull is required for pull-ups, turns and speed reductions, and push for push-overs and speed increases. Reversed control movement may result in overcontrol in pitch.

Centre of Gravity limits

Forward c.g. movement increases the nose-down moment that must be trimmed by the tailplane or elevator (elevons in case of delta wings). Thus the stick forces increase for manoeuvres and trim (fig. 4.7). Loading the aircraft to a c.g. beyond the forward limit may result in very high stick forces and manoeuvre limitations due to insufficient elevator up movements and restricted control aft movements.

Longitudinal stability decreases when the c.g. moves aft. As a result, the stick force and stick movement required for trim decrease. At the rear c.g. the aircraft may become 'touchy' and if the aft c.g. limit is exceeded the aircraft may become unstable. It may pitch up and overrotate to stall on take-off.

However, even before instability is reached an aircraft may be dangerous and difficult to control during landing in turbulence as shown by the following case.

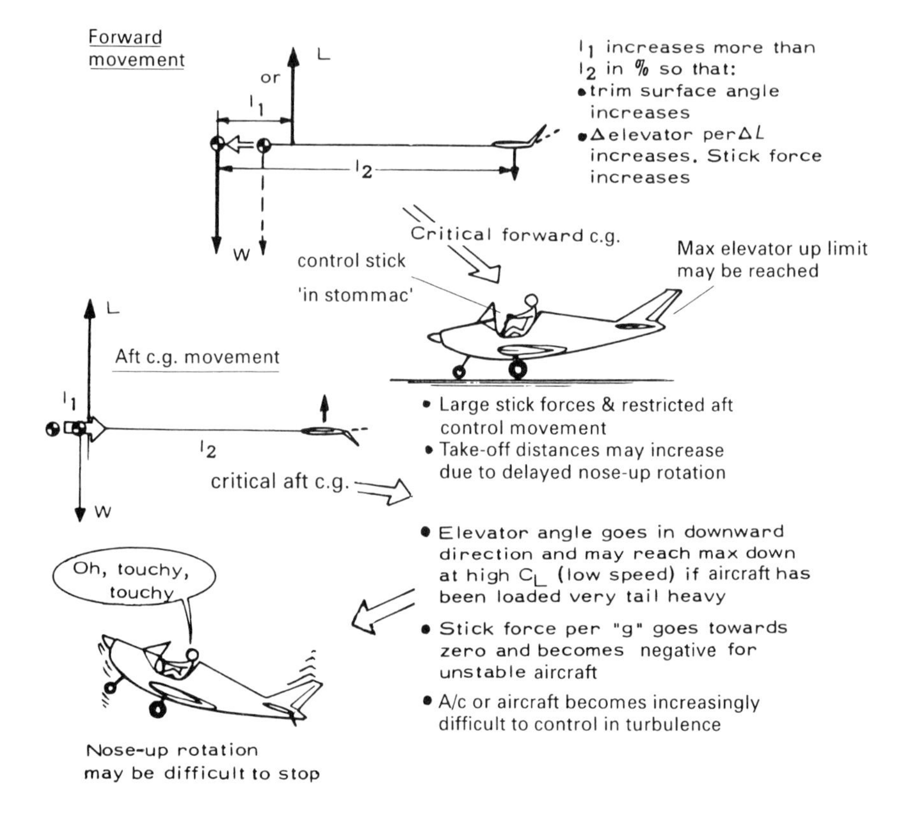

Fig. 4.7 *Centre of Gravity limits*

The crew of the ten-passenger twin-engine regional transport aircraft did not check the c.g. location before departure. Heavy passengers occupied the rear seats, the two front seats were empty, and there was no luggage in the forward compartments. Both pilots were slim. The mechanic had to press his back up against the rear of the fuselage to remove the support pole before the flight left the ramp.

No problems were reported during the take-off and flight. At the destination the air was turbulent with gusty winds. When landing flaps were selected during approach the aircraft pitched up, stalled and dived into the ground.

The use of flaps moves the wing lift resultant aft, increasing the wing pitch-down movement. This is counteracted by increased downwash behind the flaps on the tailplane. However, some pitch disturbance may remain and the effect increases with decreasing stability. The combined effect of the pitch disturbance due to lowered flaps and the pitch disturbances due to turbulence gave a nose-up pitch-rate which the crew could not stop before the aircraft had rotated to stall.

Fig. 4.8 *Overrotation and stall due to faulty nose-up trim setting*

Fig. 4.9 *Keep an eye on the c.g.!*

Roughly at the same time as this happened, a twin-engined turboprop freighter was loaded excessively tail heavy. It also pitched up and crashed on approach when the flaps were lowered.

Faulty nose-up trim setting in combination with tail-heavy loading is very dangerous and has in many cases resulted in uncontrollable pitch-ups and stalls at take-off. There may be insufficient elevator angle left to keep the aircraft nose down (fig. 4.8).

It is cheap life insurance to check the position of the c.g. and trim before take-off.

A survey of the attitudes of regional airlines after the crash showed a sad lack of awareness of the importance of loading within limits. Lack of awareness may make it easy to exceed the aft limit in general aviation aircraft which are not designed for full passenger loads, with heavy persons in the rear seats plus a full, heavy luggage load in the rear compartment (fig. 4.9).

The best advice is: whenever possible load the aircraft to a safe distance from the rear c.g. limit. For aircraft without powered controls this gives pleasant control forces and the safest flight in turbulence. A slightly rearward location of the c.g. reduces the pitch trim angles required and the trim drag.

Dynamic, Longitudinal Stability

A stable aircraft develops high frequency pitch oscillations with mainly load factor changes and long-period altitude oscillations with speed losses during climbs and speed increases during descents, so-called 'phugoid oscillations' (fig. 4.10).

Both types of oscillation should quickly dampen to zero. They are both affected by aircraft stability and increase in frequency as stability increases. The short-period frequency may become unpleasantly high at the forward c.g. limit.

At aft c.g. positions, before the aircraft becomes unstable, the phugoid motion may diverge and the aircraft may deviate from the intended path if not flown continuously. This is, of course, dangerous during low-visibility approaches.

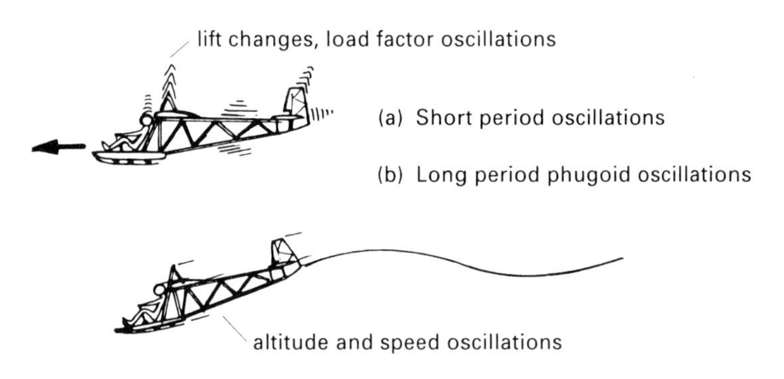

Fig. 4.10 *Pitch oscillations*

Low-speed Pitch-ups

In chapter 2, the tendency of swept-wing and delta wing aircraft to pitch up at stall was described. In these cases, even when the stall develops slowly, artificial stall warnings (stick-shakers, stick-pushers and/or warning horns) are required because if the pitch-up motion is permitted to develop, it may be impossible to stop before a full stall develops. For transports with fuselage-mounted engines pitch-up may result in a deep stall from which recovery may be impossible.

Maintain a healthy margin to swept-wing pitch-up stall. In post-maintenance tests, never flight test swept-wing transports to speeds below stall warning. Specially trained crews and special stall recovery equipment (such as spin recovery chutes) are required for this purpose! When an aircraft in deep stall yaws the fuselage vortices drift away from the tail section as shown by the smoke tests in fig. 4.11.

Fig. 4.11 *Smoke-tunnel test of yawed transport with body vortices*

This is the time to push the control column full forward to try to reduce the angle of attack below the pick-up angle if the aircraft has pitched up to a deep stall. Note that the aircraft rolls as it yaws (fig. 4.20) and falls in a yawing, rolling motion.

For aircraft with highly swept wings or delta wings, pitch-up at angles of attack of around 25° to 30° may be so sudden and violent that neither pilot nor the electronic control system can prevent pitch-up to a high angle of attack high drag flight condition from which recovery is not possible below 2 km (approximately 6500 ft) altitude above the ground due to the high sinkrate developed. Recently we had a dramatic example of this when a canard-delta pitched up out of control in a low-level tight turn during an exhibition flight. Immediate ejection and a rocket-propelled seat saved the pilot. No pilot in combat flight can keep an eye on high angle of attack pitch-up risks. The electronic control system must be designed with an inhibitor preventing pitch-up.

Small, light aircraft with swept wing-root leading edges may develop wing-root vortex flows at high angles of attack. The vortices may strike the horizontal tail and pitch the aircraft up to stall or wing failure in a stressed pull-up situation.

Transonic Tuck-under, Dives, Pull-ups and Pitch-ups

At subsonic speeds when the location of the lift resultant is fixed, the aircraft wing and fuselage pitching moment does not change with speed. In this case, as a speed-increasing disturbance increases the down-load on the tailplane or elevons and the aircraft pitches up. A speed-reducing disturbance has the opposite effect. Thus the aircraft becomes speed stable and returns to trimmed flight in a stable phugoid motion. At transonic speeds, however, the wing and fuselage lift resultant moves aft as local supersonic flow develops. When the resulting increase of the nose-down pitching moment becomes sufficiently large with increasing speed the aircraft becomes speed unstable. From trimmed level flight, it may tuck under into a transonic dive unless the pitch-control is pulled back as speed increases (fig. 4.12). Pulling out from transonic dives may be very difficult for subsonic aircraft.

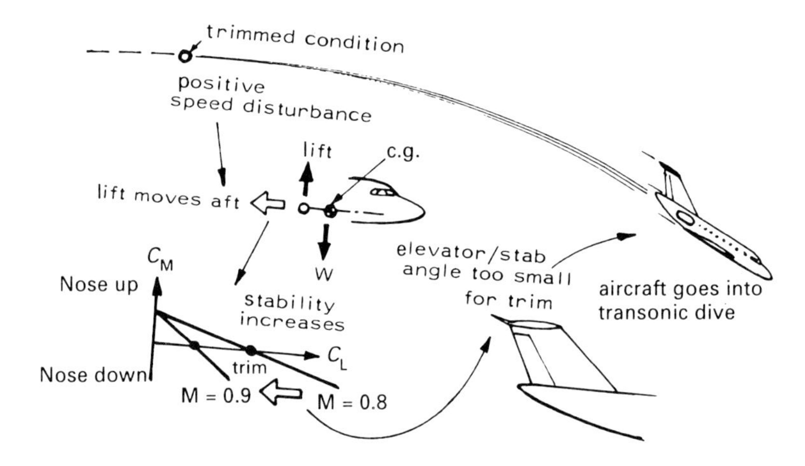

Fig. 4.12 *Transonic tuck-under*

Firstly, the increase in stability results in increased stick forces. Secondly, shock-induced flow separation may result in a large reduction of elevator effectiveness (fig. 3.4) Thirdly, the rearward movement of the lift resultant on the elevators/elevons may increase the control-surface hinge moments so much that the powered control surfaces can not be moved to sufficiently high angles for speedy pull-ups (fig. 3.12).

One way of obtaining control effectiveness for pull-up may be to retrim the tailplane (as for all flying tails on fighter aircraft). However, speedy recovery may still be impossible due to the dramatic reduction of maximum lift caused by shock-induced flow separations (fig 3.5). Transport aircraft and military aircraft with moderately swept wings, such as the one shown

Fig. 4.13 *Stall risk at transonic pull-ups*

in fig. 4.13, develop pre-stall buffeting when dived to high transonic Mach numbers but can be pulled out of the dive by throttling back and extending the air brakes.

As a result of this problem, transport aircraft have lost 6000 ft to 8000 ft in altitude during transonic tuck-under dives and executive jets have dived into the ground.

Speed reduction is a sure way of accelerating the recovery. Power reduction and air brake extension may save the day. However, some aircraft decelerate slowly and landing-gear extension may be required. This may be dangerous!

If speed is reduced with the tailplane trimmed for pull-up, the aircraft may pitch up violently (with a risk of stall or wing failure) as control efficiency is regained (fig. 4.14).

The shock-induced flow separations may also affect control in yaw and roll making manoeuvrability difficult. Reversed aileron effects (the aircraft rolls in the wrong direction) have been obtained on swept-wing fighters due to boundary layer flow separations on the wing upper surface and rearward shift of the shock wave on the lower surface when an aileron is angled downward.

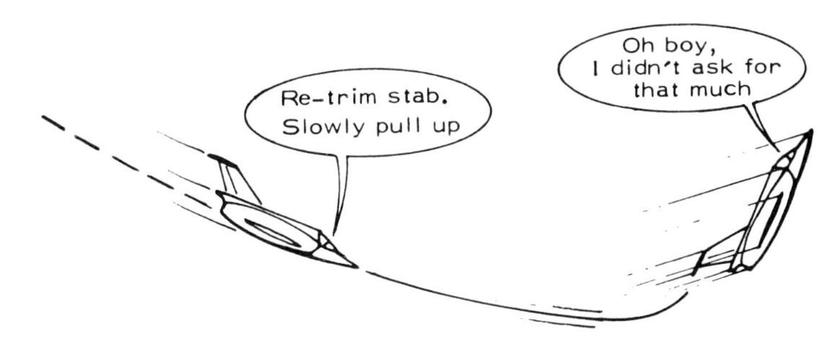

Fig. 4.14 *Pitch-up during transonic pull-ups*

Aeroelastic effects, such as twist of a lifting or stabilising surface when a control surface is moved, usually occurs at high-speed low-level flights. The result is a reduction in control effectiveness and may add to the previously mentioned transonic problems. For aircraft with fairly long wings, reversed aileron effect is obtained at the speed where the reduction of lift due to wing twist becomes larger than the effect of aileron operation. Wing weight problems may make it impossible to solve this problem other than by restricting the maximum allowable speed (fig. 4.15).

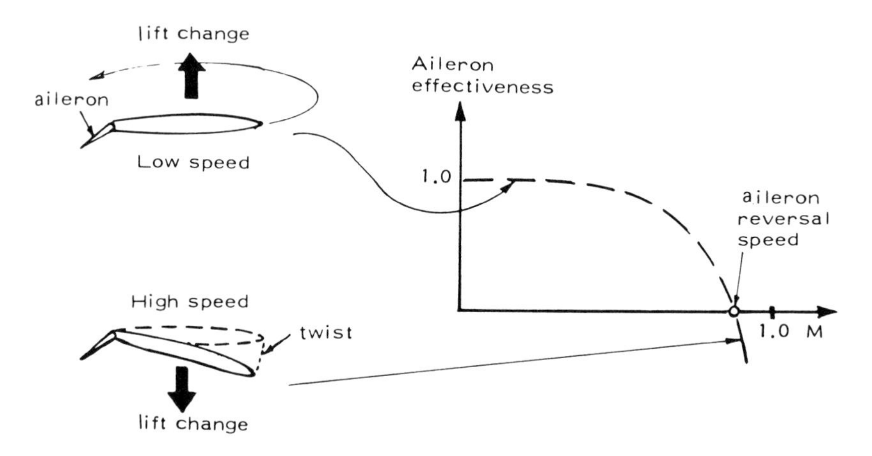

Fig. 4.15 *Aileron reversal*

Supersonic aircraft such as delta wings and aircraft with all-flying tails may have no pronounced transonic problems. One problem however, may be the change in maximum load factor available at high subsonic and low supersonic speeds. If the control surface angles are limited by large surface hinge moments at supersonic speeds, due to insufficient actuator power, the available load factors may decrease from 8 *g* below Mach one to 2 *g* above. If an aircraft accelerates into supersonic speed during a steep dive it may be impossible to pull out of it before it strikes the ground (fig. 4.16).

Intermediate angle of attack pitch-ups, caused by small transonic flow separations, may be dangerous for military aircraft in dives. Altitude or distance misjudgements may force the pilot to make an abrupt pull-up. If the pitch-up angle is reached during the pull-up the aircraft may overrotate to a high-drag angle from which recovery is not possible at low altitudes. The situation requires immediate ejection (fig. 4.17).

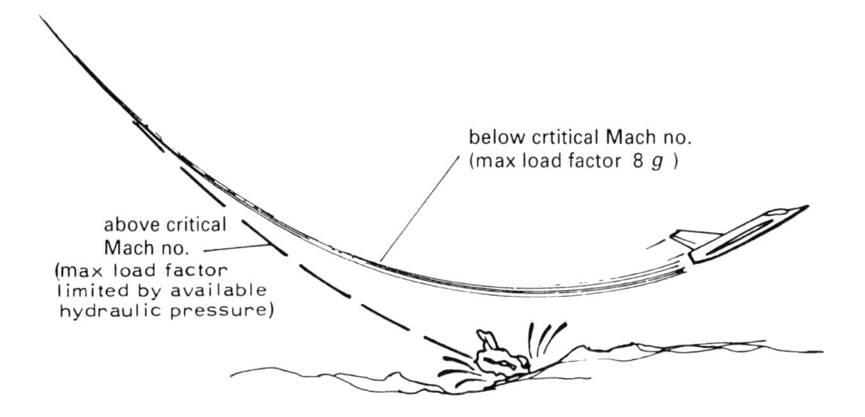

Fig. 4.16 *Pull-up from transonic and supersonic dives*

Fig. 4.17 *Pitch-up to high drag*

Pilot-induced Oscillations

Good control at low speeds requires large control surfaces. This may become a problem at high-speed low-level flights (Mach 0.8–0.9) where the large control surfaces may make the aircraft too sensitive. If the control force or the stick movement per g (i.e. the control work per g) becomes too small it is easy for the pilot to get out of phase with pitch oscillations and augment them instead of dampening them. This can result in structural failure. Increased stick work per g reduces the risk. A pilot getting into pilot-induced oscillations can get out by letting go of the stick. A stable aircraft dampens itself (fig. 4.18).

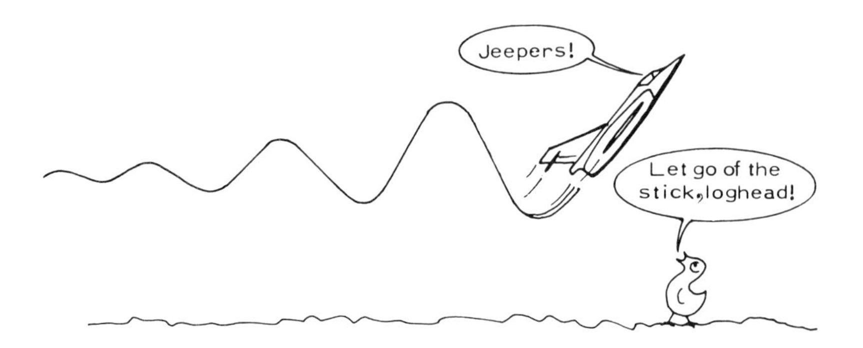

Fig. 4.18 *Pilot-induced oscillations*

Directional and Lateral Stability

The static stability requirement is repeated in yaw. For any disturbance, the aircraft must automatically return to the trimmed condition. On the other hand, too much stability may make ground handling in crosswinds difficult.

The directional control (the rudder) for single-engine aircraft may be limited by a number of factors. The most important one is, however, the control requirement during take-offs and landings in crosswinds. For multi-engine aircraft directional stability and control is usually limited by single engine failure. From a flight safety point of view this is probably the most critical directional stability situation.

The basic lateral stability requirement is that sideslips in one direction should give a rolling motion in the opposite one, i.e. left sideslip requires left stick and stick force. For straight-wing aircraft with low wings this requirement is usually met by giving the wing positive dihedral which makes the angle of attack higher for the leading wing than for the trailing wing in a sideslip.

For high-wing aircraft the fuselage side deflects the air upwards in front of the leading wing. The rolling moment thus created is sufficient to give good lateral stability and no dihedral is required.

Wing sweep increases lateral stability. Consider a swept-wing aircraft flying at a yaw angle (fig. 4.20). The unyawed span of the wing is 2b. In yaw the span of the forward moving wing increases to b_1, while the span of the trailing wing decreases to b_2. This affects the lift curve slopes of the wings as shown in fig. 4.20b, since lift is affected by wing span. The lift of the leading wing increases due to increased lift curve slope and the lift of the trailing wing decreases. As a result, the aircraft rolls away from the direction of sideslip. The yawing motion adds to this effect due to

Fig. 4.19 *Directional–lateral stability*

the increased relative speed of the forward-moving wing and the decreased speed of the rearward-moving wing.

The lateral stability of swept high-wing aircraft may become too great. Negative dihedral is often used to reduce roll due to yaw.

Since yaw induces rolling moments yaw disturbances start a combined yawing/rolling motion called a Dutch roll. It is similar to movements made by an ice skater. Due to the large effect of sweep on yaw-induced roll, the Dutch roll becomes pronounced for aircraft with large sweep angles and may have to be artificially damped.

Delta wings have special directional – lateral control problems since rudder movement changes the pressure over the wings and induces rolling

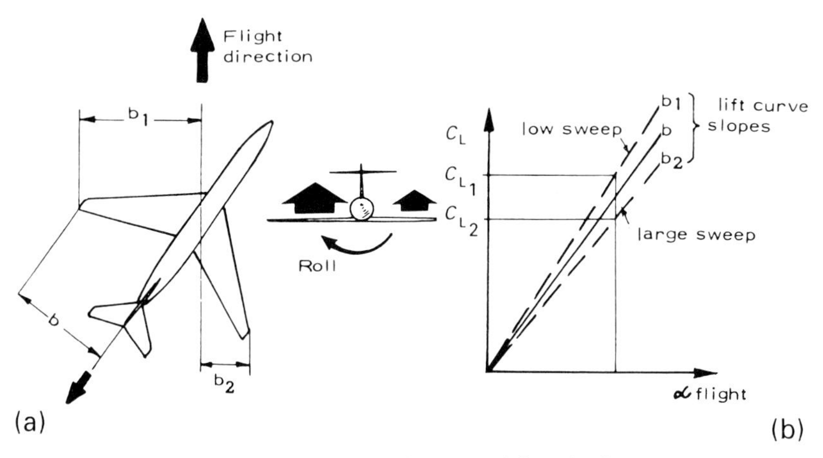

Fig. 4.20 *Effect of sweep on lift and roll*

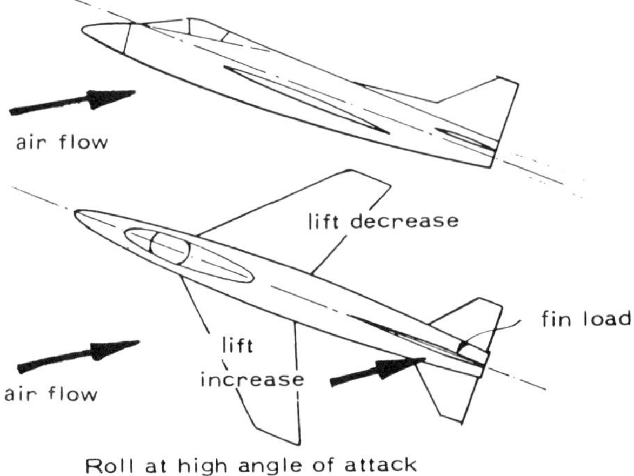

Fig. 4.21 *Sideslip (yaw) due to roll*

motions and elevon movement changes the pressure on the fin and induces yawing motions.

Long slim aircraft roll round the body axis. At large angles of attack a roll, therefore, results in a sideslip relative to the direction of flight e.g. right roll gives right sideslip (fig. 4.21).

Hence, a rolling moment, opposing the moment due to aileron operation, is obtained and the roll rate decreases until the yaw has been eliminated by the load on the fin. Control surface cross-couplings through the autopilot can be used to reduce the effects of tail–wing interferences due to control surface operation and sideslips due to roll.

Lateral control for single-engine fighter aircraft may be limited by combat roll manoeuvres, or by lateral control required to counteract roll disturbances or crosswinds in take-offs and landings. A good design rule is to require that some roll control (30%) should remain at the maximum obtainable yaw angle. This prevents the aircraft from flipping on its back in a critical yaw condition.

As for pitch, the aircraft oscillates about the c.g. even in yaw and roll. For the sake of comfort it is also required that the motions dampen to an acceptable level in a couple of oscillations.

In high-altitude high-speed flight the Dutch roll may be impossible to dampen without yaw dampers. In deep stall the Dutch roll may become extremely large due to the alternating side loads on the fin created by the body vortices.

Control Response and Damping Requirements

Aircraft must respond to control surface operation without lag and the resulting pitching, rolling and yawing motions must dampen to roughly zero at the selected trim condition. Lags and overshoots can make it nearly impossible to control an aircraft. Response lags force the pilot to make larger control inputs than those required for the desired trim change. Large countering control inputs may, thereafter, be required to stop the motions. As a result, the pilot may get out of phase with the motions and induce large oscillations round the desired trim condition. This type of pilot-induced oscillation is most critical at low speeds where aircraft damping is poor. Such oscillations may cause overrotations and aft fuselage strikes at take-offs and landings. This may result in galloping motions and crashes.

Fig. 4.22 *Delayed response and overrotation*

The problem increases with aircraft size since aircraft inertia increases faster than control surface geometry with increasing aircraft size. For this reason, artificial response boost and damping through the autopilots may be required for large aircraft. When the pilot makes a normal control input the autopilot exaggerates the input to accelerate the motion and thereafter dampens the motion with opposing control surface movement.

Transport aircraft with fuselage-mounted engines and long fuselages (long forebodies) may be difficult to control in gusty crosswinds on slippery runways. A sudden cross-wind gust may set up a yawing motion resulting in a sideways skid sending the aircraft off the runway.

Engine Effects on Stability

For all propeller aircraft the corkscrew rotation of the slipstream may cause some trim changes as power is changed. Furthermore, flow velocity

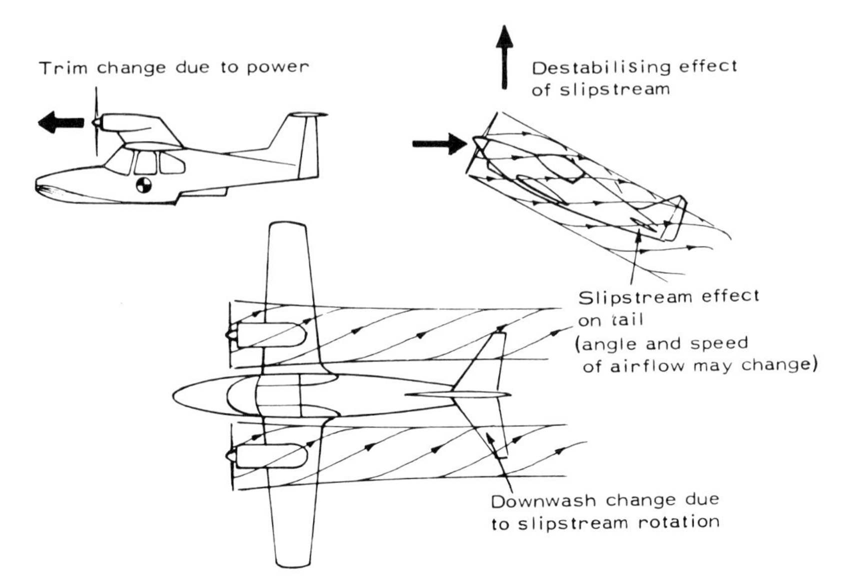

Fig. 4.23 *Typical power effects on trim*

Fig. 4.24 *Jet thrust has lift component*

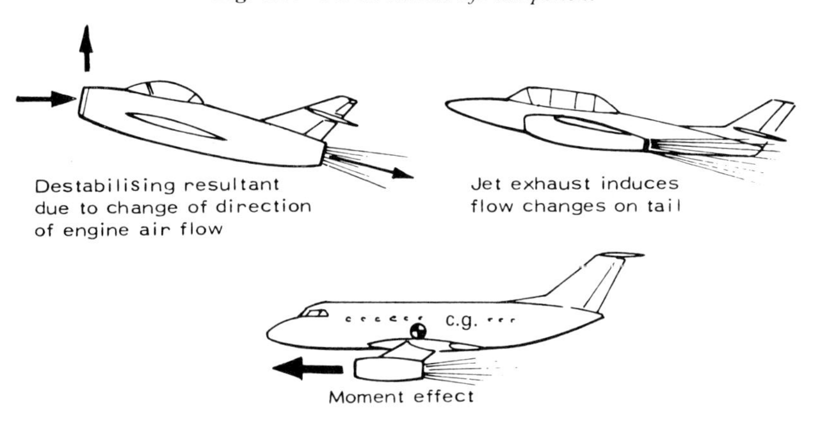

Fig. 4.25 Jet effects on stability and trim

changes over the tail surfaces due to power changes also affect trim (fig. 4.23). The figure shows trim changes due to engine location being distant from the centre of gravity, the destabilising effect of slipstream deflection at the propeller, and trim changes due to slipstream rotation, a change of downwash and a change of flow velocity at the tail surfaces.

The jet engine effects on lift are not as large as the propeller effects. However, at large angles of attack in landing, the lift component of the engine thrust is notable (fig. 4.24).

There are, however, other jet effects on stability and trim (fig. 4.25). These are similar to the effects on propeller aircraft, namely a destabilising force due to engine air deflection, changes in flow over the tail surface due to nearby jet exhaust, and moment changes due to jet thrust at a distance from the aircraft's centre of gravity.

However, the fact that wing mounted engines give pitching moments when thrust is changed has been used by a clever airline crew to fly a heavy transport, which had a complete loss of control system hydraulic pressure, down to a crash-landing which many passengers survived.

For multi-engine aircraft the critical condition for yaw and roll control is usually the single engine failure case. Since slipstream deflection over the wing and flaps gives considerable lift, single engine failure immediately creates a large rolling moment towards the dead engine. This is augmented by the roll due to yaw. The result may be a quick roll which is difficult to stop.

When a propeller operates in an upwash in front of a wing, the angle of

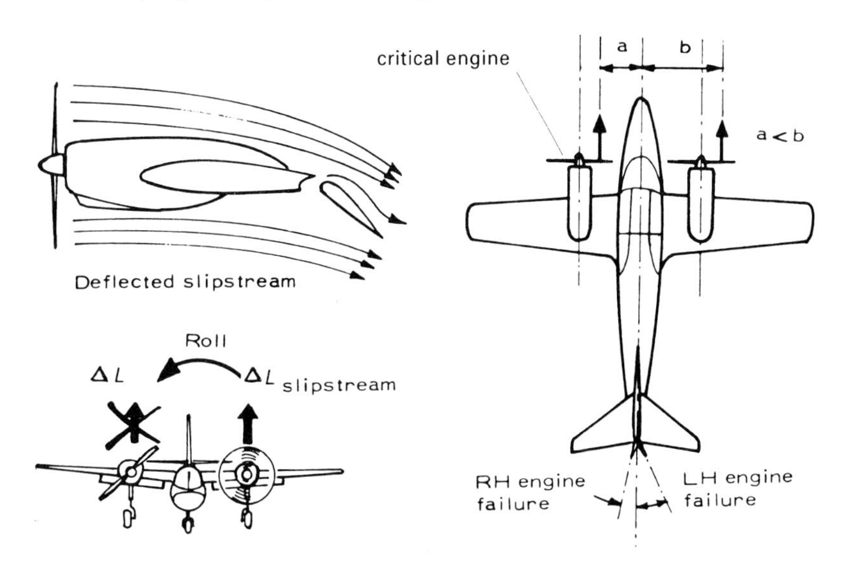

Fig. 4.26 *Loss of lift and resulting roll due to engine failure in front of flap*

attack of the down-going blade rotating into the upwash is larger than the angle of the blade rotating upwards with the upwash. This moves the resulting thrustline of the propeller towards the downwards-rotating blade. For an aircraft with two clockwise rotating propellers (looking forward) the right engine, therefore, has the largest yawing moment. This makes the left engine the 'critical' engine in case of single engine failure (fig. 4.26).

The minimum control speed with one failed engine is flight tested with the dead-engine propeller feathered and the aircraft banked 3° towards the running engine. The bank gives a sideslip and a side-load on the fin which helps maintaining yaw control and flight course. Landing-gear and flaps are retracted. The aircraft is new and the crew well trained. They know what is going to happen.

Compare this to a sudden, unexpected engine failure during take-off in less than perfect weather conditions. Control may be lost and the aircraft may flip on its back at speeds well above the flight-tested minimum control speed. The loss of lift on one wing combined with roll due to yaw give a rolling motion that cannot be stopped by full opposing aileron only. To recover it is vital to:

- kick immediately full rudder towards the running engine and follow up fast with full roll opposing aileron
- keep the nose down to maintain speed
- feather the dead engine
- retract the landing-gear
- increase power on the running engine
- retract flaps as soon as speed permits.

The following shows how fast control can be lost. The student pilot had throttled back one engine and was decelerating towards V-MCA with the 'good engine' at low power. Gear and flaps were retracted. Near the minimum control speed, the instructor told the student: 'You are doing O.K., add power.' By mistake the student pushed the good engine's throttle full forward. 'We were in an inverted spin before I could think,' the student said afterwards.

The accident pictured in fig. 4.27 is typical of uncontrolled roll due to engine failure. The aircraft was on a low-level approach to an airport when the pilot was told by air traffic control to turn right. The pilot added power and turned. As he did this, the right engine quit. The aircraft continued rolling and struck a hill in a 90° bank.

Single engine failure of the outboard engine of a four-engine transport, may be very critical even if the failure occurs above the minimum control speed.

Fig. 4.27 *Loss of roll control due to single engine failure*

The problem here is the angular yaw acceleration which may be so large that it may be impossible for the pilot to counteract with the rudder before the aircraft has yawed to an angle where roll control is lost (approximately 15° yaw for some aircraft). This may happen even if it is possible to trim the asymmetric thrust at a safe angle of yaw in tests. Large overshoots of the trimmed angles may come as the result of delayed actions. One reason for delay may be the large pedal movement needed to reach maximum rudder angle (never approached in normal operation). This problem caused so many training accidents that flight training was discontinued.

Even twin-engine transports with fuselage-mounted engines have had problems with single engine failures mainly because pilots have kicked the rudder in the wrong direction.

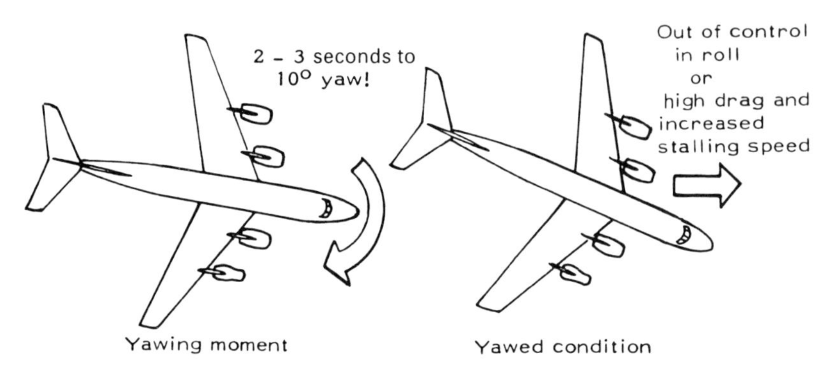

Fig. 4.28 *Outboard engine failure*

Spin

At stall, the change in drag with angle of attack is large. Therefore, if an angle of attack difference is obtained between the two wing halves at or near stall, for example through a rolling motion or a gust, considerable difference in drag between the left and right wings may be obtained. Asymmetric stall may lead to roll due to difference in lift, and to yaw due to drag difference.

The aircraft falls in a yawing – rolling motion (fig. 4.29). This is spin. The

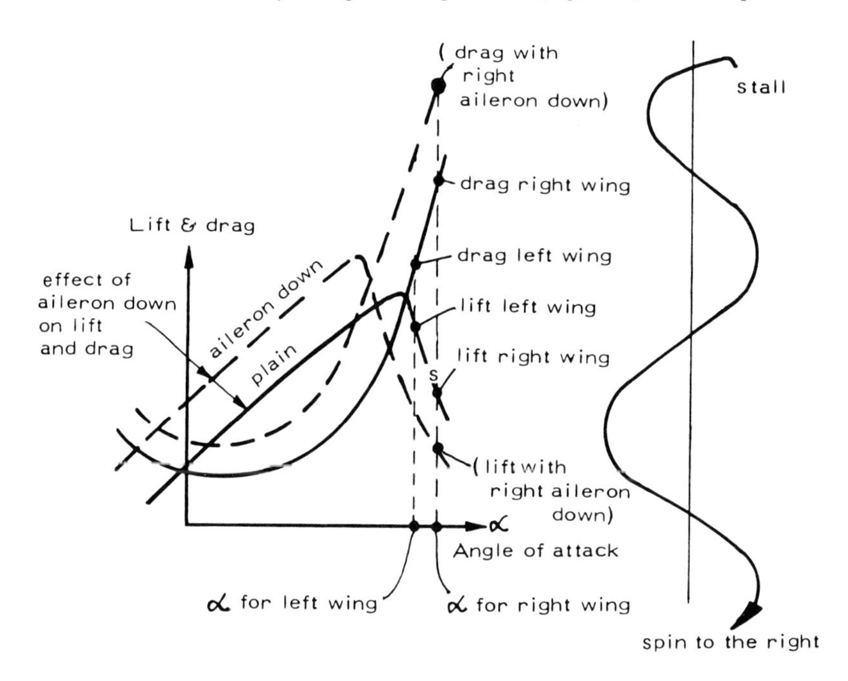

Fig. 4.29 *Spin*

motion will not stop by itself since the down-going wing in a spin also has the largest angle of attack and thus, the lowest lift and the highest drag. The motion itself maintains the difference in forces needed to keep the spin going.

Opposing aileron operation will not stop the rotation since, as shown in chapter 2, the movement of a control surface or flap lowers the stall angle of a wing. Thus, aileron operation may further exacerbate the stall of the down-going wing in a spin. The reduction in stall angle caused by aileron movement makes this manoeuvre dangerous, even near stall. Opposing a rolling motion with aileron at stall may trigger asymmetric stall and spin.

Stall strips can be used at the outboard wing leading edge to slow down the spin rate of an aircraft. This is dangerous. It increases the risk of wing-tip stall and sudden departure into spin at stall. Stall behaviour can be improved by adding stall strips in the wing-root sections.

Most light aircraft recover from spin if the yawing motion is opposed by use of the rudder, ailerons are neutralised and the stick is moved forward to lower the angle of attack. However, for high-speed military aircraft spin recovery may be a complex manoeuvre requiring careful analysis and training. Aircraft with poor control surface efficiency in spin and good response below stall may flip from spinning in one direction to spinning in the opposite direction or from normal to inverted spin during recovery. The following advice may be helpful.

- *General aviation* Stay away from spin but be sure to learn the spin characteristics of your aircraft and how to recover.

- *Transport aircraft* Stay away from spin! Recovery may require anti-spin chutes! However, if you are stuck in a deep stall, observe that the aircraft Dutch rolls due to alternating sidewash by the body vortices as the aircraft yaws. The vortex downwash on the tailplane is at a minimum at maximum yaw angle; in a sideslip, the vortices drift away from the empennage (fig. 4.11). This is the time to push the control column full forward to try to pitch the aircraft downwards before the vortices again exert maximum downwash on the tailplane. Spoilers on the fuselage top could destroy the vortices but have not yet been used.

- *Delta wings* If you are stuck in a superstall with a delta wing or canard–delta you will find that the aircraft oscillates in pitch between roughly 30° and 90°. During pitch-down, slam the stick forward and give maximum thrust in order to increase the pitch-down rate and get on the low angle of attack side of the pitching moment 'hump'. Slow down the pitch-down rate immediately after recovery or you may be thrown into an inverted superstall.

Porpoising and Bouncing

Amphibious aircraft taking off from or landing on water may begin to porpoise. A wave striking the bow may give the aircraft a sudden pitch-up to lift-off angle of attack. If the pilot overreacts the aircraft may strike the water surface nose first and pitch up again. If the pilot does not immediately throttle back and keep the stick still in a landing position, the porpoising diverges (pilot-induced divergence) and finally the aircraft dives into the water (fig. 4.30).

Fig. 4.30 *The final porpoise dive*

A high-speed landing bounce caused by flare misjudgement or a dive for the deck when landing on a short runway may lead to pilot-induced 'porpoise-leaps' down the runway (caused initially by too much nose-down control after the first bounce). If the landing-gear can take the repeated 'crashes' the high-speed bounce may slowly change into a low-speed bounce with subsequent loss of control. The best solution to a high-speed bounce may be full power, increasing speed and to go round again. However, in some cases the aircraft may be too badly damaged to go round again safely.

However, whether going round again or continuing with the landing, it is important to ease the control forward from the far aft landing position in order to prevent speed loss in the air after the bounce.

Fig. 4.31 *Porpoising after hard, high-speed landing*

Divergent bounces with loss of control have happened to transports with fuselage-mounted engines and the c.g. at or behind the aft limit. If the c.g. at touchdown is located behind the main-gear station, an uncontrollable pitch-up may be obtained. Low-speed, high angle of attack landing bounces with nose-up rotation may result in stall-induced pitch-up with loss of pitch and roll control. This has happened to empty transports with fuel in the outboard wing tanks (fig. 4.32).

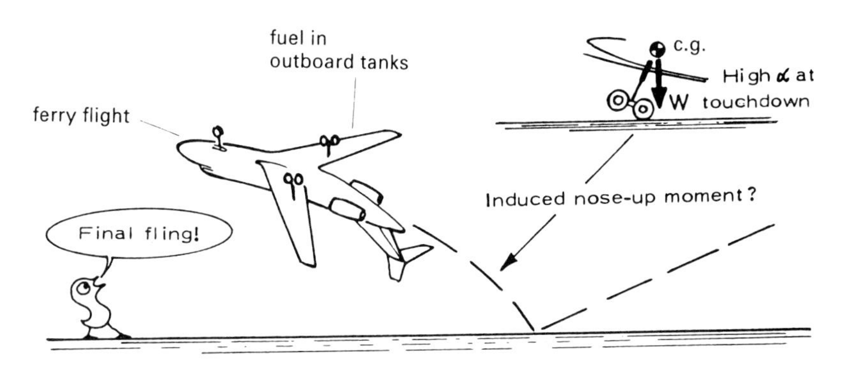

Fig. 4.32 *Bounce-induced pitch-up*

Aircraft with high inertia relative to the aerodynamic forces may have control problems during low-speed flares. Thus, jet transports with fuselage-mounted engines and long, stretched fuselages may have a tendency to overrotate due to high inertia and poor damping in pitch (fig. 4.22). After-body strikes on landing have been reported.

For aircraft with low stall angles with flaps lowered (aircraft with effective flaps but no slats) there is a certain risk of rolling on the nosewheel in high-speed landings (fig. 4.33). Since the aircraft now rotates round the nose-gear in yaw it is weathercock stable and yaws off the runway even in

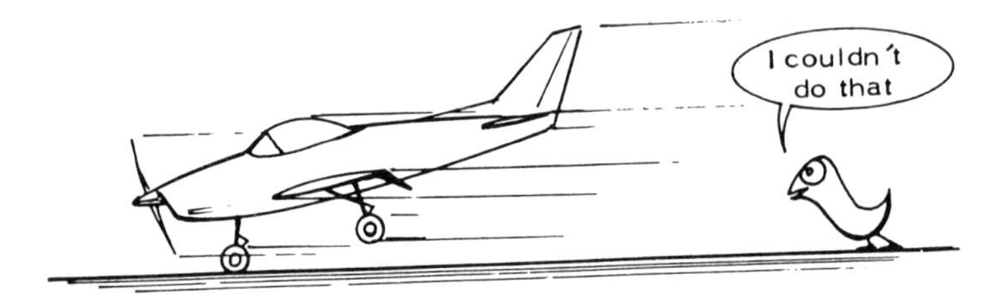

Fig. 4.33 *Wheelbarrowing*

slight crosswinds. Sidewise motion as the mainwheels touch down may roll the aircraft on to its back.

Every aircraft has its own stability and control characteristics. Good knowledge of these and of the effects of operational conditions on the handling qualities are essential for safe flight.

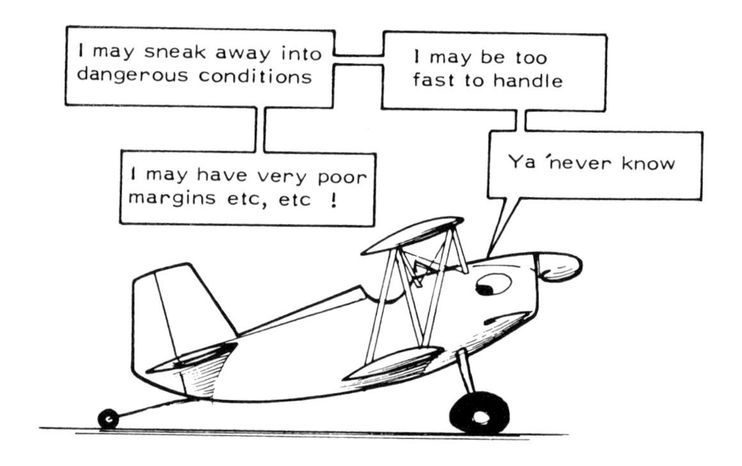

Fig. 4.34 *Stability and control characteristics vary*

CHAPTER 5
LOST REFERENCES

The most common accident caused by loss of external references is that which occurs during flight into instrument meteorological conditions (I.M.C.) weather by visual flight rules (V.F.R.) pilots with 'gethomeritis', pressing along in deteriorating weather. A pilot not trained and experienced in instrument flying may lose control in a few seconds when entering a cloud (fig. 5.1). The lucky ones regain control if the height above the ground is sufficient when they come spinning or diving from the cloud base. The unlucky ones fly into the ground or shed the wings in pull-ups when trying to avoid a crash. In these situations, it does not help very much if the aircraft has good handling qualities. Manual control requires both visual references and seat-of-the pants feeling.

Transition to instrument flying requires some time and should be made while the pilot (crew) still have external references. It is difficult to catch up with moving instruments in time to avoid a critical situation with no reference for guidance. The following accident illustrates this. The pilot had a night-flight permit and was used to night flying. However, he was not instrument rated.

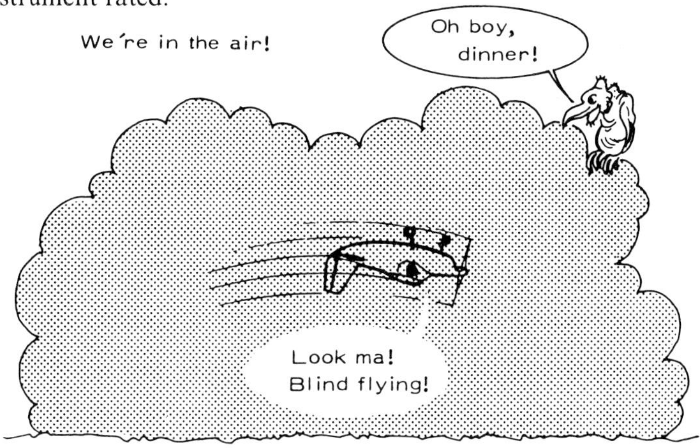

Fig. 5.1 *Disorientation in clouds*

The weather at the airport and in the runway direction was clear with stars shining. The pilot was cleared for a right turn to a navigation fix after take-off. The track passed over open fields and forests. There was no light on the ground. Unknown to the pilot, low clouds had drifted in towards the right side of the airport. When the pilot turned, therefore, he suddenly flew into a black 'bag' with clouds above and a completely dark surface below. He managed to maintain speed and course but failed to check his altitude. The result is shown in fig. 5.2.

Fig. 5.2 *Crash due to sudden loss of external references after take-off*

Judging from all the accidents I have investigated, I find that professional pilots know the difference between V.F.R. and I.F.R. (instrument flight rules) flight and realise the need to be fully established on instruments before transition from V.M.C. (visual meteorological conditions) to I.M.C.

However, not all pilots realise that what appears to be V.M.C. actually is I.M.C. where flying on external references is extremely dangerous. Some of these V.F.R. traps, where external references are poor or change rapidly, are discussed here. Typical V.F.R. traps during landings are shown in fig. 5.3.

Even in clear weather you may have to make your approaches in conditions where it may be difficult to judge the height above the ground using only external references. Conditions may rapidly deteriorate when haze, fog, rain or snow reduces visibility. The professional pilot must be aware of the need to use available instruments under all conditions. This is especially true when you have to land on runways without V.A.S.I.s,

Fig. 5.3 *V.F.R. traps*

without approach lighting or with the approach lights mounted along the ground instead of on poles along a proper approach path.

Mirror Lake – Brother's Cottage

A pilot notified A.T.C. 'I am descending to low altitude over the lake to take a look at my brother's cottage'. It was one of these beautiful summer days with sunshine, warm weather and not a cloud above the mirror-smooth lake surface. The descent ended with a big splash which left an expensive amphibian at the lake bottom and an expensive patient at the local hospital.

Now, it has been proved repeatedly in accident after accident that it is not possible to judge the height above a mirror-smooth lake surface even if the visibility is infinite. But this is basic and a well-trained commercial pilot would never make such a mistake – or would he?

Fig. 5.4 *Diving through the mirror surface*

Black Holes

A four-engine jet transport was approaching an airport located at the outskirts of a small city at the edge of an uninhabitated mountain region. The night was crystal clear and the airport could easily be seen at the edge of the dark forest. So, the crew elected to make a V.F.R. approach. During the approach, the aircraft descended below the glide slope and hit a peak in the approach zone several miles from the runway.

Investigation of the accident showed that there had been several incidents where crews had descended below peak height, had lost sight of the airport and had shouted to A.T.C: 'Why did you switch off the runway lights?' Air traffic control responded by shouting 'Pull up, you have descended below peak height'. After the accident a light was installed on the peak.

As a result of this and similar accidents, a simulation of V.F.R. flights towards airports located at the edges of black surfaces was made. Of twelve

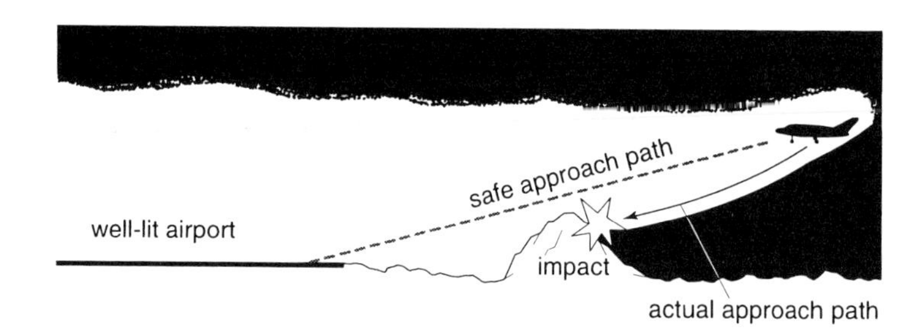

Fig. 5.5 *Descending into a 'black hole'*

well-trained airline crews selected to fly the simulated V.F.R. approaches over the 'black holes' eleven crashed before reaching the runway threshold!

Both the simulations and the accidents show that there is no safe way of making V.F.R. approaches to airports located in areas where you do not have good altitude references.

White-out

Winter with sunshine, crisp air, clear weather and no turbulence is an inviting time for a flight. Winter is, unfortunately, also the time when we have repeated reports of aircraft getting flipped on to their backs when pilots touch down in the snow in front of the threshold.

For the same reason as it is difficult to judge the height above a mirror-smooth lake surface or a 'black hole' it is difficult to estimate the height above smooth snow-covered ground. For this reason, pilots approaching runways surrounded by contourless snowfields must be sure to check the glide slope nearly as carefully as when making an instrument approach, even in clear weather.

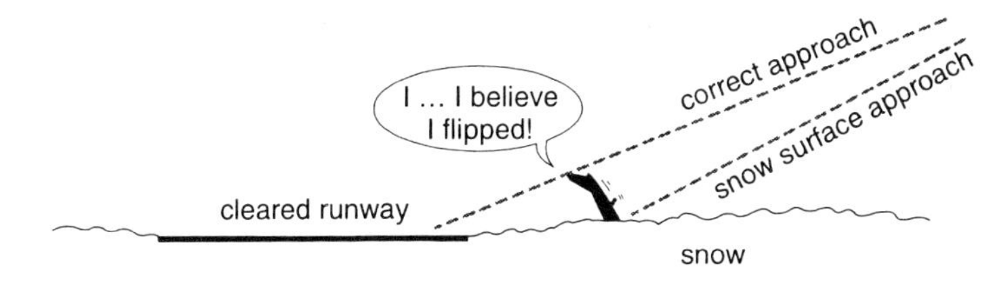

Fig. 5.6 *White-out*

Haze, fog, snow, runway slope, length, width and the shape of the surrounding terrain add to the problem of making approaches to 'winter airports'. What appears to be V.M.C. conditions may in changing weather conditions rapidly become difficult I.F.R. approaches. In contourless 'white-out' conditions, where fog or falling snow blends with the white ground, you have no way of knowing what is up and what is down.

Helicopter pilots taking off or landing on snow-covered surfaces may find themselves sitting inside a white 'dome' of snow blown up by the rotor. Loss of control is nearly immediate.

Duck-under

A pilot approaching an airport with low-level fog patches drifting over the surface runs a special risk, irrespective of whether the approach is made

V.F.R. or I.F.R. Assume that the pilot (the crew) has established visual contact with the runway at low altitude and, in connection with this, has completely abandoned instrument flight. If a fog patch suddenly drifts in over the approach zone it becomes nearly 'natural' for the pilot to push the control forward in order to maintain contact with the ground. The pilot tries to 'duck-under' the fog.

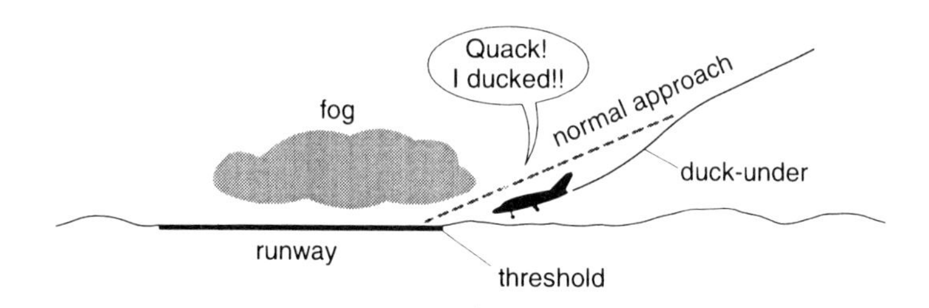

Fig. 5.7 *Duck-under*

Many accidents caused by the irresistible temptation to duck-under show that if you do not have good approach-sharing responsibilities (one pilot always on instruments) and abort in time under these condition the risk is fairly great that you will slam the aircraft into the ground near the threshold in a duck-under manoeuvre. Drifting fog patches suddenly steal your references and your immediate reaction is to start chasing them.

Landing Illusions

Investigations show that a pilot's ability to make safe approaches is affected by the terrain surrounding the airport and by the runway length, width and slope. As a result of this, you may be misled even in daytime.

Observe the following.

- Up-sloping terrain in front of a runway gives you an impression of being too high. You run the risk of descending below the glide slope and touching down before reaching the runway (fig. 5.8). An up-sloping runway gives you the same impression of being too high (fig. 5.9).

- Down-sloping terrain in front of the runway gives you the impression of being low (fig 5.10). And a down-sloping runway will give you the same impression of being below the glide slope. You run the risk of touching down too far down the runway.

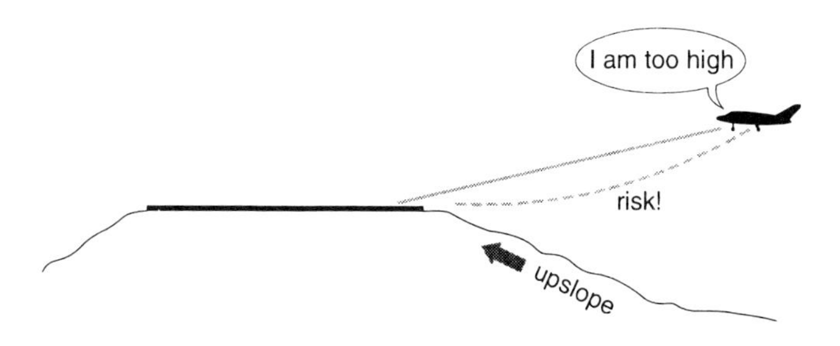

Fig. 5.8 *Illusion created by up-sloping terrain*

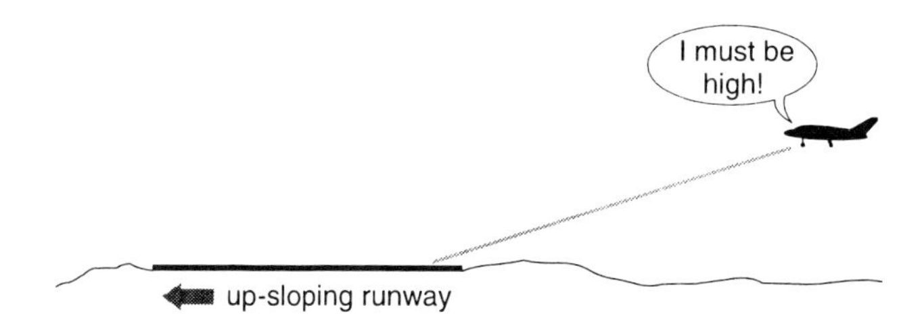

Fig. 5.9 *Illusion created by up-sloping runway*

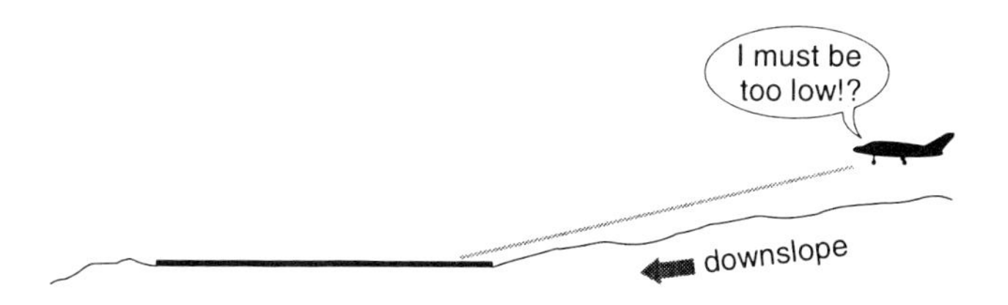

Fig. 5.10 *Illusion created by down-sloping terrain*

- The runway width and length have various effects on the pilot's height perception. A short and wide runway may give you the impression of being too low, while a long and narrow runway may give you the impression of being too high. Both depend to some extent on what you are used to (fig. 5.11).

The landing illusions add to your problems when making 'black hole' approaches. Various combinations of runway slopes, lengths and widths will make the approach strictly an I.L.S. affair. In daytime, especially in

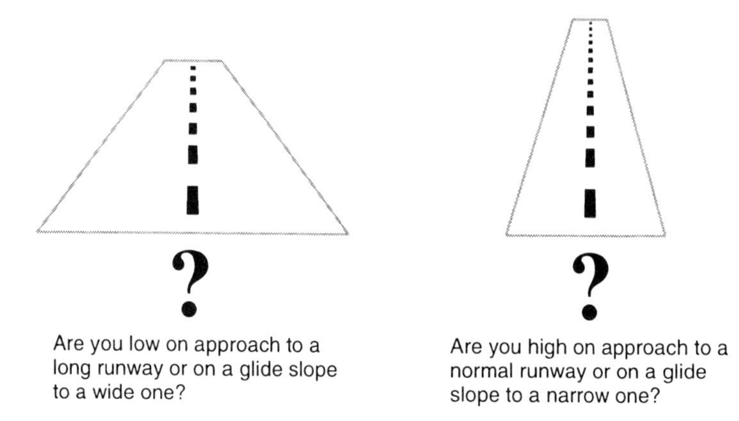

Fig. 5.11 *Runway length and width illusions*

conditions with reduced visibility (haze, rain on the windshield, etc.), the combination of terrain and runway features may also trap you and set you up for a crash near the threshold.

The following example shows how easy it is to make a mistake. In order to save money, the airport authorities in a certain country decided to paint the centreline stripes on new regional airports half as wide as usual (fig. 5.12). As a result of this, the domestic airline had a number of damaged nose-gears to fix as the pilots flared too late during the approach. The centreline widths have now been increased to normal size. The cracked nose-gears resulting from this 'money-saving' action show that even well-trained airline crews can misjudge heights when subjected to unusual references.

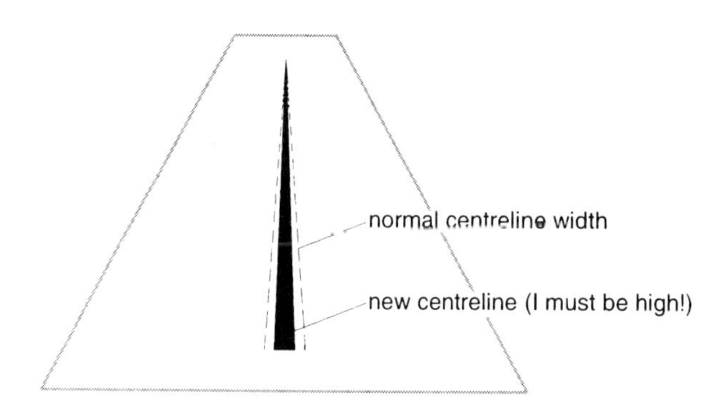

Fig. 5.12 *Narrow centreline gives wrong height impression*

The pilot flying the crashed aircraft show in fig. 5.13 ran into deteriorating weather on a night training flight, and decided to immediately make a 180° turn back to the airport which was still clearly visible in the distance. During the turn over a dark forest, the aircraft lost height and collided with trees just as the pilot began to pull up. It was a typical 'black hole' accident.

Fig. 5.13 *Collision with a dark forest during 180° turn at night*

CHAPTER 6

AIR LOADS – FATIGUE AND OVERLOADS

Aircraft are subjected to a large number of varying loads. These loads are static and dynamic loads due to flight, and loads due to ground operations (fig. 6.1). In spite of this, it is possible to design aircraft that can operate safely in severe, hostile environments most of the time.

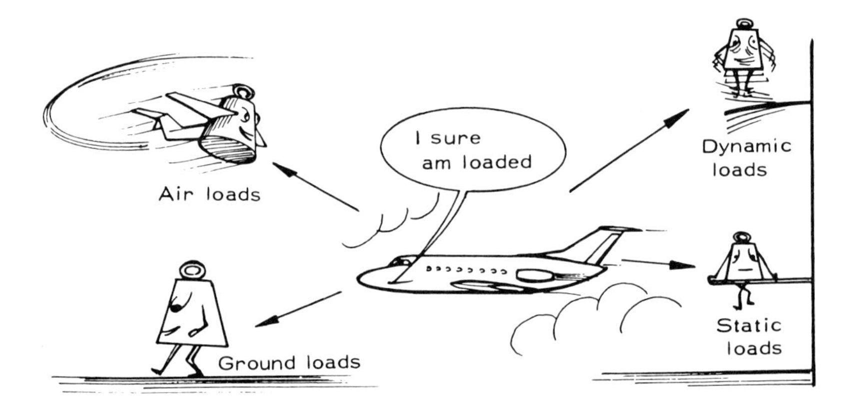

Fig. 6.1 *Aircraft loads*

The Design Environment

For every aircraft a design environment is stipulated by the authorities and the manufacturers. Maximum static design loads are chosen for various aircraft categories and the dynamic loadings are estimated from the operational environment. From these a design load spectrum is determined (fig. 6.2). The load spectrum shows the frequency of loadings from the maximum design loads estimated to occur only once in the aircraft's design life to loads of increasing frequency occurring in every flight.

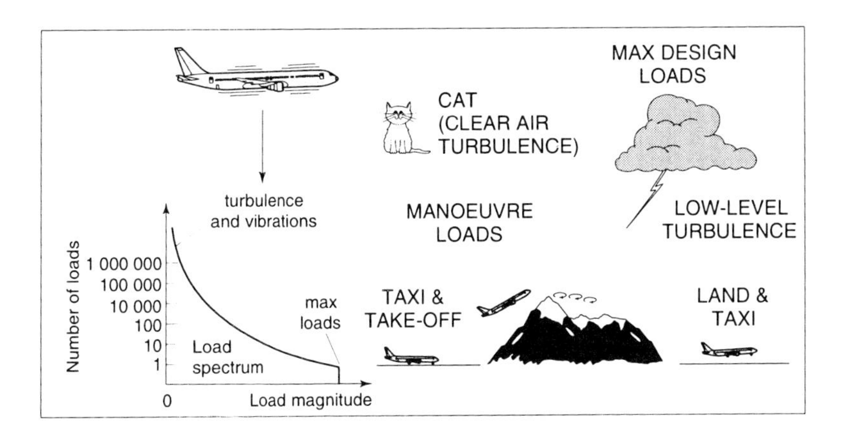

Fig. 6.2 *Operational environment*

Aircraft are designed for three types of loading:

1. maximum static loads
2. fatigue loads – repeated loads
3. flutter – aeroelastic loads

Maximum Static Wing Loads

The following maximum positive and negative design load factors (n_z) apply:

Fighter/attack aircraft	+ 8	to	–3
Aerobatic aircraft	+ 6	to	–3
Utility aircraft	+ 4.4	to	–1.8
General aviation aircraft	+ 3.8	to	–1.5
Transport aircraft	+ 2.5	to	–1.0

These factors give the maximum allowable wing loads, the load limit. They are multiplied by a safety factor of 1.5 to give the ultimate loads, the wing failure loads. The load factor limit may be presented as a function of speed in a V–n diagram and as a function of Mach number and altitude in a flight envelope (fig. 6.3).

The V–n diagram shows how the available load factor at stall increases with the square of the speed until at certain speeds (V_1) the load factor limit is reached. At approximately 20% higher speed (V_2) the wing fails at stall.

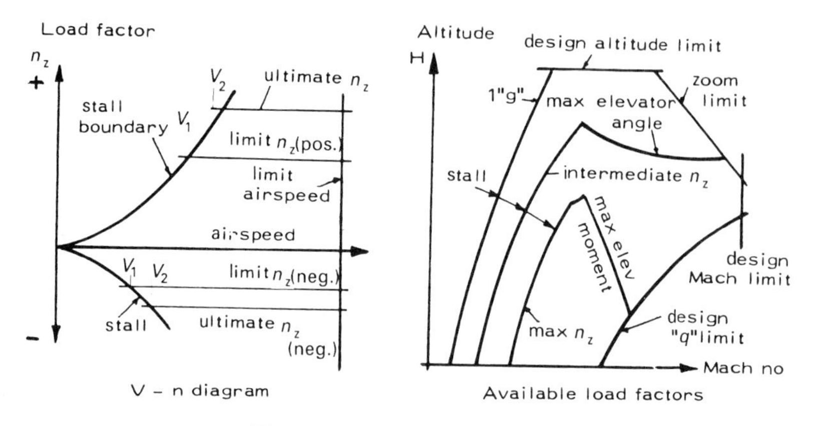

Fig. 6.3 *Available load factors*

The altitude–Mach diagram shows the combinations of Mach numbers and altitudes where the maximum load factors and other design limits can be reached, all of which are used for structural design.

The reliability of the main load determinations, such as the spanwise and chordwise distribution of wing loads (fig. 6.4), is very good, due to wind-tunnel testing and modern calculation methods. The problem that causes accidents is that the loads in many cases can be exceeded. Pulling high load factors with aircraft not designed for aerobatics has resulted in many wing failures, such as the one shown in fig. 6.5. Flights into severe updraughts

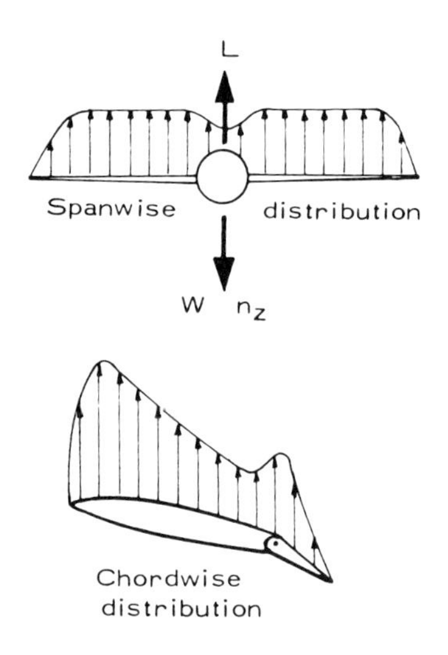

Fig. 6.4 *Wing load distribution*

Fig. 6.5 *Wing failure in pull-up*

and downdraughts at speeds where wings may fail before they stall have also caused structural failure accidents (*see* chapter 7, 'Dangerous Winds').

Moreover, the spanwise pressure distribution is affected by aircraft roll. In roll the down-moving wing gets the extra roll damping load shown in fig. 6.6. This loads up the outboard wing section thus moving the lift resultant outboard, increasing the wing-bending moment in the root section.

If at the same time the aileron is moved downward to stop the rolling

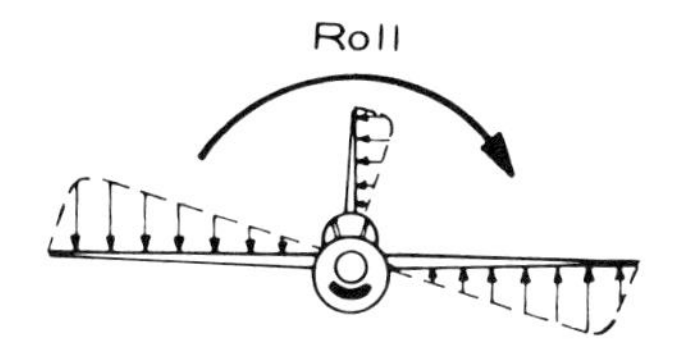

Fig. 6.6 *Damping loads due to rolling motion*

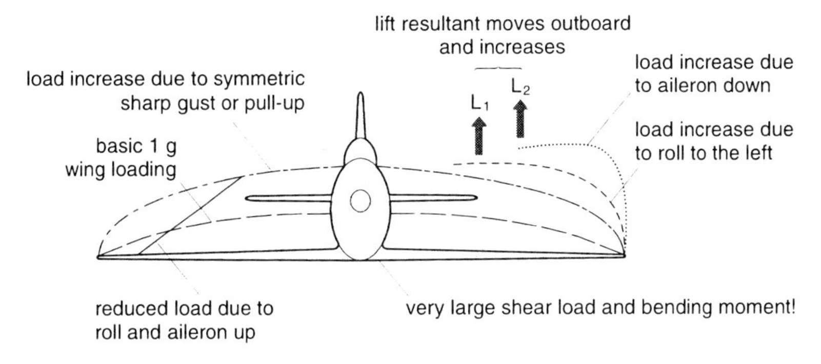

Fig. 6.7 *Spanwise loading due to pull-up, roll and aileron movement*

motion, the loading of the outboard wing section increases and may become critical in combination with pull-ups or flights into severe updraughts at speeds when wings fail before they stall (fig. 6.7). Aerobatic roll and pull-up manoeuvres with general aviation aircraft may, for this reason, be very dangerous.

Wing-mounted external stores such as tip-tanks reduce the load on the wings in pull-ups due to the downward-directed *g* loads on the tanks, when they are full (fig. 6.8). However, the tanks give endplate effects on the wings, moving the lift resultant outboard and increasing the wing-bending moment. Empty tanks may, therefore, increase the wing-bending moments and increase the risk of wing overload.

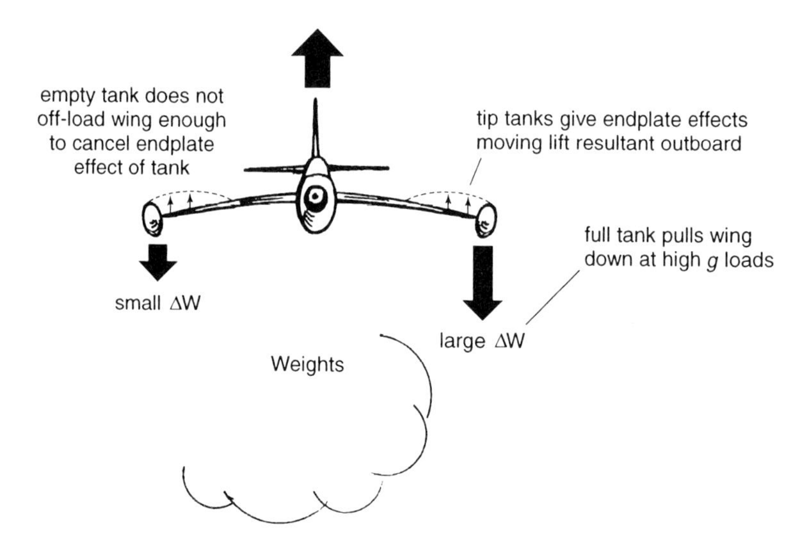

Fig. 6.8 *Effect of tip-tanks on wing load*

Empennage Loads

The maximum side force on the fin is often determined by the maximum pedal force the pilot can apply (fig. 6.9), or by trim loads in the case of a failed wing-mounted engine of a multi-engine aircraft. But for modern, slim, supersonic aircraft the fact that the aircraft does not always roll along the flight axis, so that roll at high angles of attack may lead to excessively high tail loads must also be considered (fig. 6.10).

It has also been known for heavy aircraft to have the fin blown off in horizontal gusts in mountain terrain.

Yaw due to asymmetric thrust (in flight reversal of one engine) may overload the fin when the autopilot or crew try to stop the yaw with full

Fig. 6.9 *Maximum pedal force*

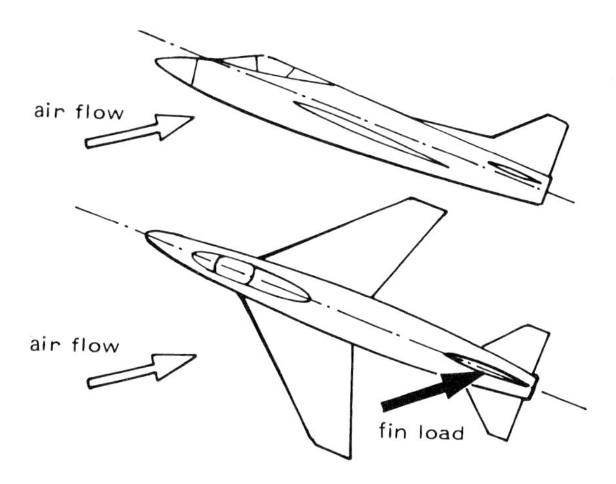

Fig. 6.10 *Effect of roll on tail load*

opposing rudder. Also, the large yawing, rolling and pitching motion, caused by asymmetric thrust and yaw dampener hard-overs, may result in total in-flight disintegration of a transport aircraft. This happened to a twin-engine jet transport when one engine reversed during a high-altitude climb.

Tailplane loads may be determined by trim loads under conditions giving maximum nose-down moments of the wing and fuselage, i.e. maximum forward c.g. at maximum load factor in pull-ups (fig. 6.11).

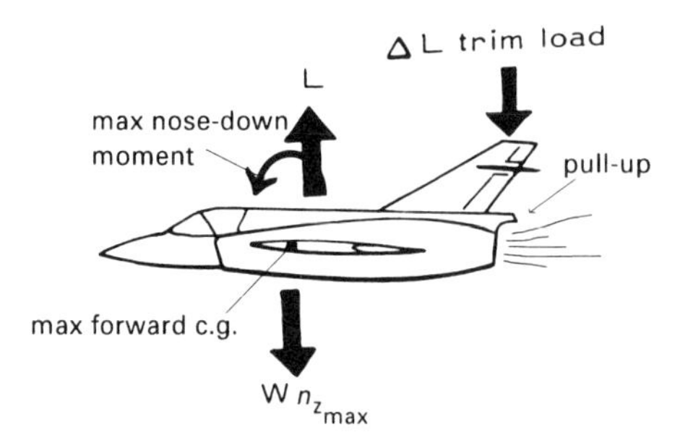

Fig. 6.11 *Maximum trim-load on tailplane*

Fuselage Loads

The loads on fuselage skins of aircraft with pressurised cabins are determined by the differences between cabin pressure and outside pressure at maximum flight altitudes. At this altitude the pressure differences may give loads around 6 m^{-2} (fig. 6.12). In spite of this, cabins do not fail in static pressure overloads. Overload failures are caused by terrorist bombs or by bird strikes (fig. 6.13). The figure shows the result of an osprey collision during a departure climb.

Fuselages may also be overloaded in hard landings (fig. 6.14), when large

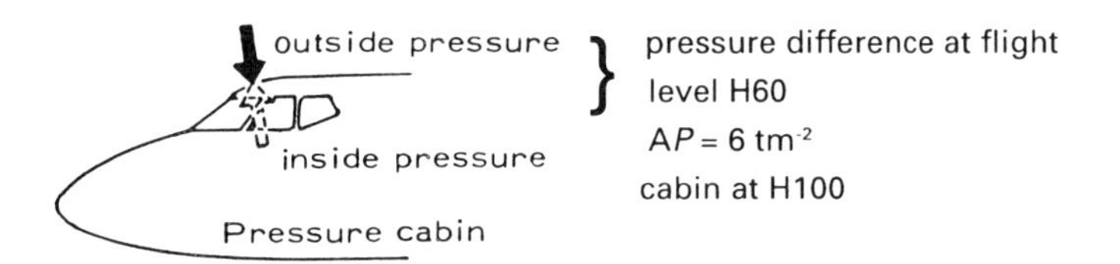

Fig. 6.12 *Pressurised cabin loads*

Fig. 6.13 *Bird strikes may knock holes in cabins*

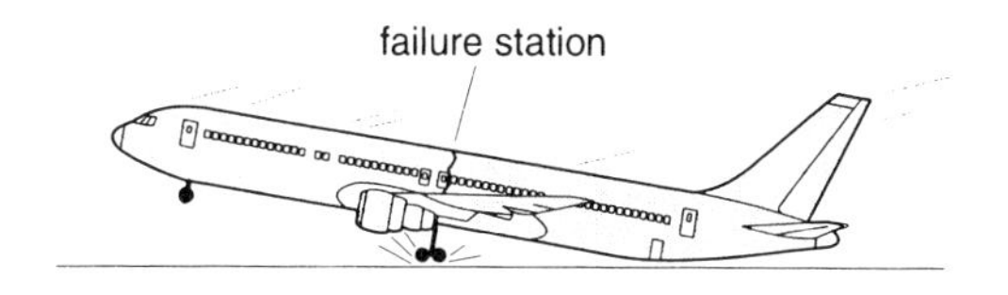

Fig. 6.14 *Hard-landing loads on fuselage*

downward-directed mass forces on the forward and aft fuselage sections result in fuselage failure near the main-gear station. Large landing loads may also overload the wings in downward bending.

Landing-gear Overloads and Fires

Landing-gear may be subjected to compression overloads in hard landings. However, at touchdown large bending loads are imposed while the wheels spin up (fig. 6.15). Hard landings at high speeds may, therefore, be critical. However, loads imposed on landing-gear running down in airport ditches or soft ground are probably the most common cause of landing-gear overload failures.

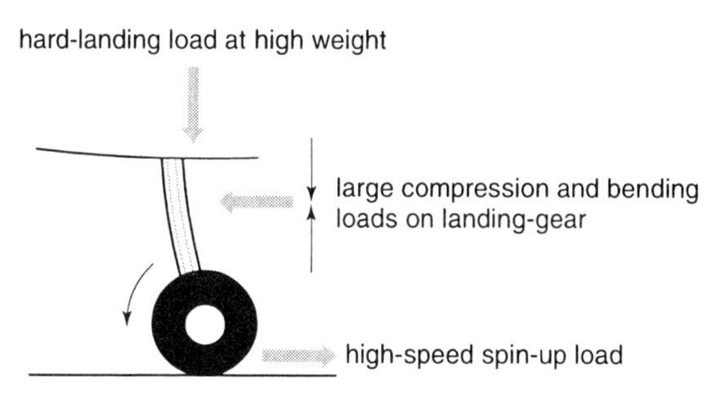

Fig. 6.15 *Landing and spin-up loads may combine to overload landing-gear*

The heat that can be absorbed by landing-gear brakes is limited. Heat stored in the brakes in cases of excessive braking may start landing-gear and tyre fires or may heat up the tyres of a retracted gear after lift-off sufficiently to cause a tyre explosion resulting in wing structural failure. Tyre temperatures of transport aircraft must be checked (for instance after aborted take-offs) and the brakes must be allowed to cool down before braking or taking off again (fig. 6.16). High tyre heat may also be caused by underinflation, long taxi distances, high taxi speeds, sharp taxi turns,

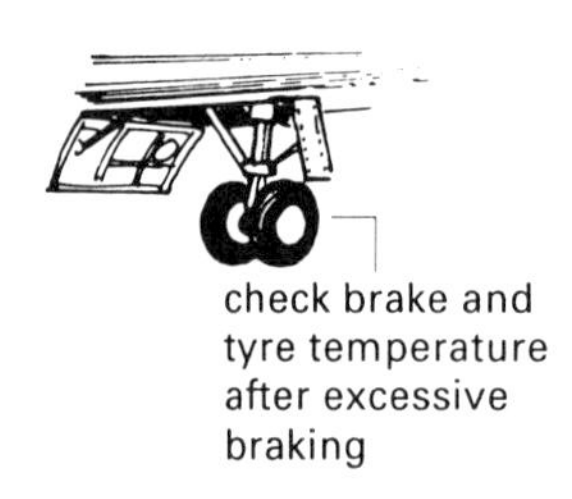

Fig. 6.16 *Risk of brake fire and tyre explosion*

high speeds in turns, and exceeding the maximum rotational speed limit of the tyres.

Antiskid systems may not be able to control braking on runways with slick spots quickly enough to prevent the wheels from locking on the most slippery spots. When the locked wheels skid on to coarse runway surfaces the risk of tyre failure becomes very great.

Tyre failure at high speed with large take-off weights is still a serious cause of transport aircraft accidents. With a failed tyre, braking efficiency is reduced and it may be impossible to stop the aircraft before the runway end is reached. Damaged tyres usually fail at critically high speeds when the loads on the tyres are greatest. Underinflated tyres are dangerous even if they are not damaged. Traction waves in these tyres may heat the tyres sufficiently to cause rubber separation or tyre explosion. When tyres are paired it may be difficult to see if one tyre is underinflated. The underinflated tyre increases the load on the correctly inflated tyre in a pair. Both tyres may fail.

For these reasons it is very important to check the tyres for both damage and correct pressures before take-off.

Loads on Engine Installations and External Stores

Pod-mounted engines and stores mounted under swept-back wings may be subjected to high loads in the spanwise direction, in combination with the *g* loads during pull-ups, due to the outboard airflow below swept-back wings (fig. 6.17a).

Combinations of pull-ups and rolls add inertia loads on the pod-mounted engines or stores. Together with the air loads this has resulted in overloads (fig. 6.17b).

Fuselage-mounted engine pods may be overloaded in hard landings. The following case illustrates what may happen. The captain extended the lift dampers of a DC-9 by mistake just before touchdown. Increased drag and

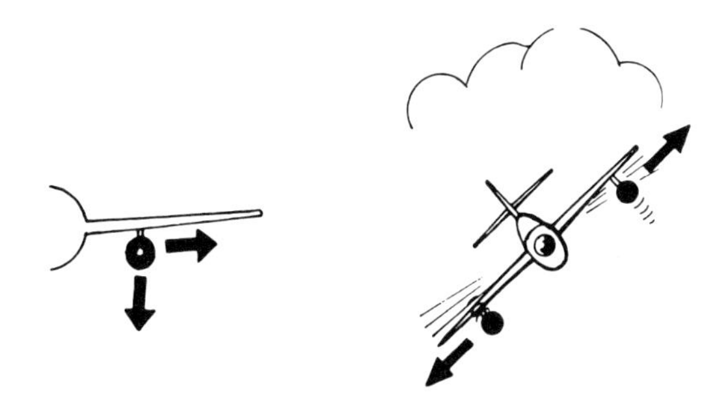

(a) Loads in pull-ups (b) Load increase due to roll

Fig. 6.17 *Loads on external stores and engines in pull-ups and rolls*

decreased lift slammed the aircraft on to the runway. Inspection after the landing showed the following damage (fig. 6.18).

- Both engines pods were twisted downward.
- The wings were overstressed in downward bending.
- The fuel tanks were leaking.
- The main-gear was damaged.
- The fuselage was overstressed in downward bending.

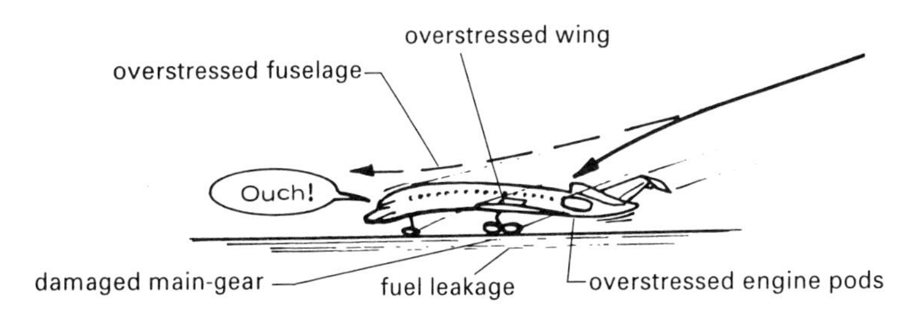

Fig. 6.18 *Effects of premature use of lift damper*

Cabin Floor Failures

Cabin floors on pressurised jet transports are not designed to take the large downward-directed floor loads obtained due to sudden losses of pressurisation in compartments below the floor caused by high-altitude failure of large cargo doors. Blow-down doors in the cabin floor are designed to

Fig. 6.19 *Failed cargo door*

relieve the load on the floor in case of pressure losses below but are not sufficient to take care of very sudden, total pressure loss in the cargo hold. When the cargo door on the aircraft in fig. 6.19 blew out, eight seats in the passenger cabin fell through the floor and the passengers were sucked out of the opening.

In-flight Impacts

A jet transport climbing out from Stockholm, Arland Airport, struck something with a wing. Inspection after landing showed that it had hit a fish! Now, fishes do not fly above Sweden. Obviously the aircraft had scared the food out of the mouth of a seagull. The wing leading edge was dented.

Bird collisions may knock out front windows on any aircraft if the speed is high enough. Fig. 6.20 shows the damage inflicted on a fighter aircraft's nose at a very low bird-strike angle at high-speed low-level flight. Bird impacts may also seriously damage wing leading edges.

Fig. 6.20 *Bird-strike on fighter nose*

Flights below the anvil of thunderstorms may result in collision with very large hailstones (fig. 6.21). The hailstones may knock front windshields white, reducing forward visibility, or seriously damage wing leading edges (fig. 6.22).

Fig. 6.21 *Collision with hailstones in clean air below the anvil of a thunderstorm*

Fig. 6.22 *Hailstone damage to wing leading edge*

Lightning Strikes

Lightning strikes do not usually damage metallic structures but have in a very few cases caused fuel tank explosions. Never refuel during thunderstorms! Always ground an aircraft before refuelling. Static electricity sparking between a refuelling nozzle and a fuel tank may cause an explosion.

Composite structures with high electrical resistance may heat up and delaminate when struck by lightning (*see* helicopter rotor in fig. 12.24). As the use of composite materials increases, the risk of structural damage due to lightning strikes may become serious unless metallic electricity leads are built into the structure in order to prevent structural overheating from poor electrical conductivity.

As fly-by-wire and electronic control systems increase, the risk for serious damage to control systems and antennas by lightning strikes increases.

Consequences of Overloads

All aircraft may be overloaded in manoeuvres, hard landings and severe turbulence without immediately visible structural damage. However, overloads may cause permanent structural deformations which weaken the structures and later may make them fail at loads far below the design load limit. Overloaded structures must be inspected and repaired.

Walk-around Inspection

A careful, preflight walk-around inspection may be one of your best lifesavers. These inspections have revealed damaged structures, damaged tyres, open hatches, covered static pressure ports and pitot tubes, leaking fuel, leaking oil, etc. Walk-around inspections cost little but may save much.

Fatigue

All metal structures develop cracks when subjected to a large number of load changes even if the loads are small. By means of the load spectrum (fig. 6.2 and fig. 6.23) the designer determines the load variations throughout the design life of the aircraft.

Cracks begin to develop at stress concentration points such as sharp corners. The designers avoid stress concentration in order to minimise crack development. Vital structural parts of expensive aircraft are fatigue-tested in rigs where the fatigue loadings based on the load spectrum are accelerated. By means of theories, tests and structural check-ups, the fatigue life of the aircraft and the need for crack inspection, crack-propagation checks and repairs are determined.

However, the load spectrum is based on operational estimates of take-off and landing frequencies, manoeuvre distributions, flight times and altitudes, etc., and on atmospheric turbulence models for flights at different altitudes. The design load spectrum can be exceeded without ever exceeding the maximum design load (fig. 6.23).

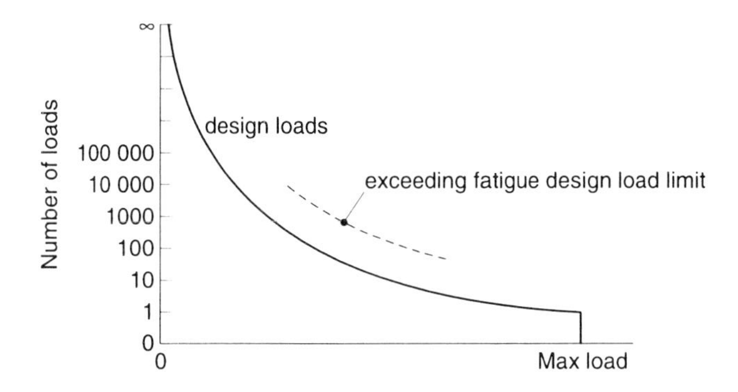

Fig. 6.23 *Exceeding the design load spectrum*

Flying in environments that are much more severe than the design environment used for the determination of the design load spectrum may have dramatic effects on the fatigue life of an aircraft.

Take-off and Landing Loads

Continued operation from rough airport surfaces may shorten the fatigue life of an aircraft, especially its landing-gear (fig. 6.24). The landing-gear of light aircraft have failed in fatigue when the shock-absorbing wheels

Fig. 6.24 *Rough field operations shorten fatigue life*

have been replaced by skis and take-off and landing runs were made on rough snow and ice surfaces.

Fatigue life is also shortened by frequent take-offs at maximum weight especially if the take-off speed is increased due to wing contamination. It is also shortened by frequent hard landings (turbulence, maximum landing weight, short fields, etc.), and by high rates of hard braking (short fields, high altitudes and high-speed taxying followed by hard braking to turn) fig. 6.25. The following accident is a good example of what may happen.

The fully loaded charter transport had landed at an airport in an underdeveloped country for refuelling. This was a regular charter refuelling stop. To reduce ground time the crew used high-speed taxying and hard braking. On the way to the departure runway with maximum take-off weight, the crew braked hard at the turn-off from the taxiway to the runway. During the turn, the right main-gear failed, the fuel tank ruptured and fire broke out. There were many fatalities. The accident had all the ingredients of reduced landing-gear fatigue life.

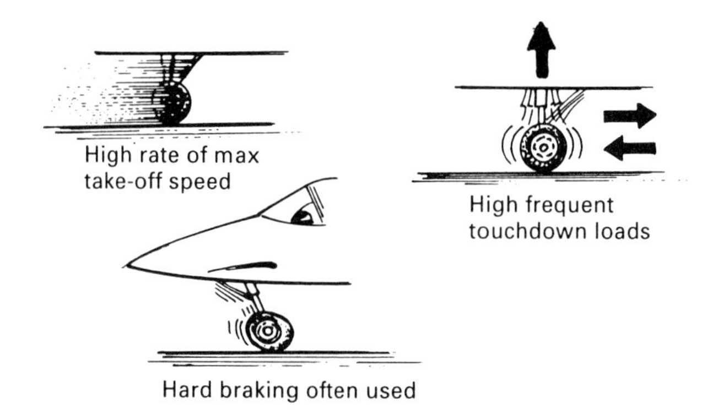

Fig. 6.25 *Landing-gear fatigue due to continuous high landing-gear loadings*

Effect of Atmospheric Turbulence on Fatigue Life

The turbulence level is higher at low altitudes than at high (fig. 6.26). The fatigue life of an aircraft designed for high-altitude flights may, therefore, be reduced by 50% if it is used regularly for low-altitude flights.

However, while low-altitude turbulence may be absent on cool days above flat terrain, the gust intensity may be severe in mountain terrain. Regular flying in turbulent environments wears out aircraft more quickly than flying in calm conditions and requires special checks to avoid unexpected, fatal structural failures.

Fig. 6.26 *The effect of altitude on turbulence level*

Pressure Cabin Fatigue Failure Risks

High-altitude fatigue failure of a pressurised fuselage usually results in total fuselage disintegration. The risk of fatigue crack development increases with the number of cabin pressure load changes, i.e. with the high number of climbs and descents associated with short-range flights and the increased gust intensities at reduced cruise altitudes.

The results of a pressure cabin fatigue failure of an aircraft flown by a short-range operator is shown in fig. 6.27.

Fig. 6.27 *Pressurised cabin fatigue failure*

Flap Fatigue Risks

The airflow over the flaps becomes increasingly more turbulent as the flap angles are increased. At large angles the flow separations result in load

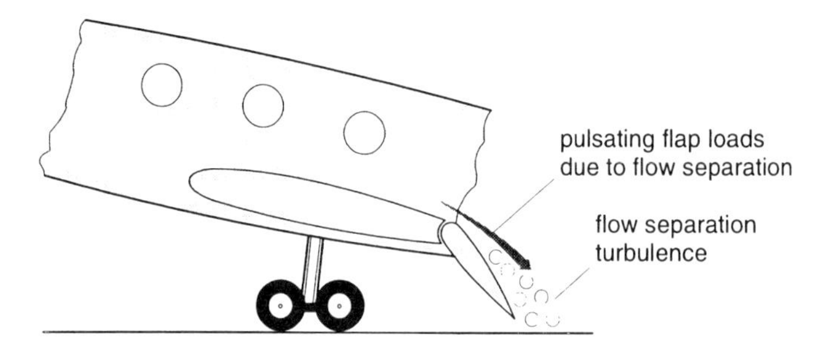

Fig. 6.28 *The flap region is a high-risk area for fatigue failures*

variations which shake the flaps. The pulsating loads induce crack development and, after some time, fatigue failure. This is taken into consideration when designing the flaps and developing inspection routines and maintenance schedules. In spite of this the flap region remains a high-risk area for fatigue damage and failure (fig. 6.28).

The fatigue risks increase with increasing flap use (short-range flights), increasing flight time with extended flaps (early establishment of stabilised approaches) and repeated selection of high flap angles at high approach speeds.

Note that if flaps are lowered at speeds above the flap limit speed the following may happen:

- the flaps may be overloaded and fail
- the flaps may be overloaded and deformed; cracks may begin to develop in the overloaded structure
- the flap nose-down pitching moment may send the aircraft into a dive.

Engine Installation Fatigue Risks

The vibration level in the immediate vicinity of a reciprocating engine is usually high (fig. 6.29a). The level increases with increasing propeller wear and damage. This is an area with risk of fatigue failure.

The loads at podded jet engine installations are complex (fig. 6.29b). The pod and pylon are subjected to high forward and rearward engine thrust as well as thrust reversal loads, vertical loads in turbulence, take-offs and landings, and side-load variations due to side gusts and yaw oscillations. On top of this, there are the gyroscopic reaction loads of the engine. This is a high risk area for crack development and fatigue failure unless the pod installations are carefully inspected and maintained.

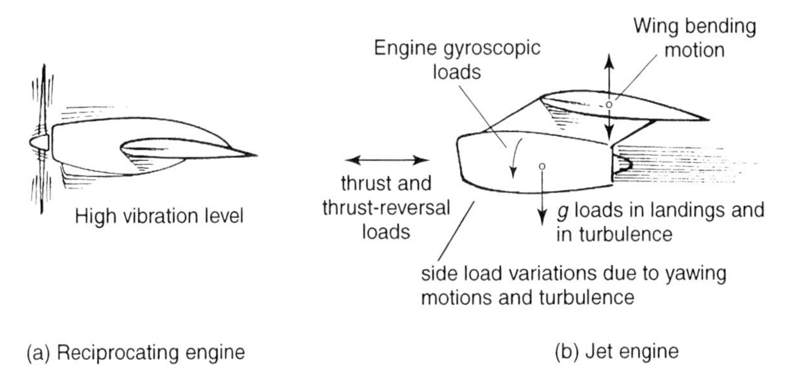

Fig. 6.29 *Fatigue-inducing vibrations and loads at engine installations*

Pods and pylons seldom fail but when they do the engine may move forward due to residual thrust and rip off the wing leading edge sections, increasing the stalling speed of the aircraft, making landing difficult. On a four-engine aircraft the inboard engine may, in case of failure, move outboard and knock off the outboard engine. As a result, both engines and large sections of the wing leading edges on one wing may be lost, resulting in a dramatic increase in stalling speed of the damaged wing.

Crews flying aircraft with pylon-mounted engine pods should note that a sudden loss on one or two engines may indicate that the engines have fallen off and taken the sections of the wing leading edge with them. When this happens, the damaged wing may stall at twice the normal speed. Do not slow down to the published approach and landing speeds. It is better to maintain flight speed down to landing than to lose roll control and dive in to the ground. Remember the DC-10 at Chicago which lost one engine, and the Boeing 747 of Amsterdam which lost engines three and four. Both crashed but could have been landed had high speed been maintained.

Vortex Flow Induced Vibrations

Vortices shed by the wings of transport and fighter aircraft and fighter aircraft noses at high angles of attack (figs. 2.30 and 2.31) may strike the tailplane and fin and induce large vibration loads or load up the surfaces to high whiplash loads which may knock off the elevator mass balance and the pitot tube on the fin.

Oscillation and Buffeting Vibration

Aeroplanes in flight oscillate in pitch, roll and yaw. The oscillations create load variations which increase with increasing pitch, roll and yaw in turbulence (fig. 6.30).

Furthermore, artificial damping is sometimes required to keep oscillations about the trimmed condition at acceptable, low levels. The damping motions of the control surfaces combined with control inputs and aircraft oscillations introduce continuous load changes (fig. 6.31).

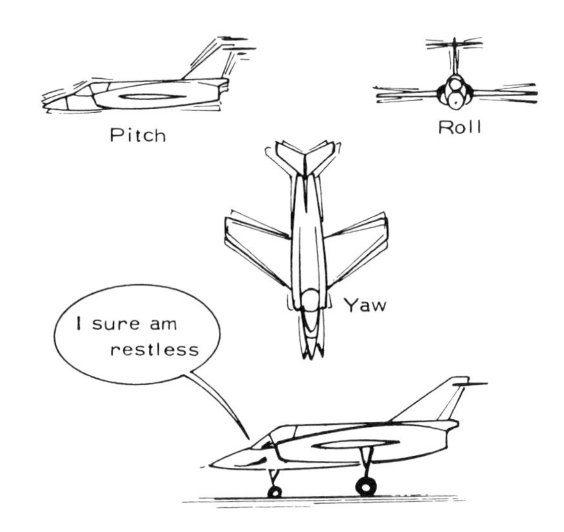

Fig. 6.30 *Oscillations introduce load changes*

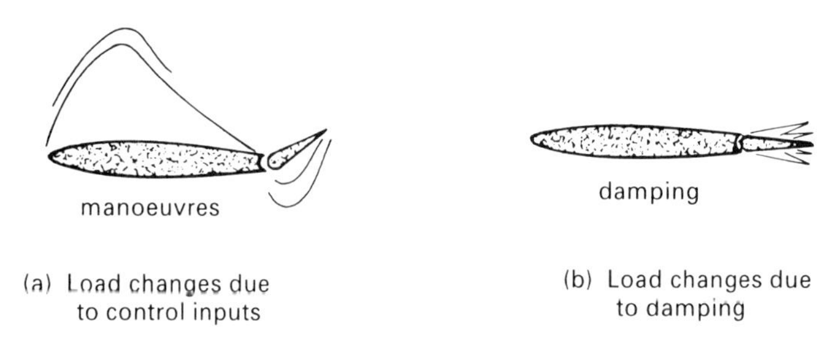

Fig. 6.31 *Local load-change wear on control surfaces due to small control surface movements and damping*

At high angles of attack near-stall, large buffeting load variations may be obtained (fig. 6.32). Fighter aircraft repeatedly flown at near-stall buffeting in combat turns develop wing fatigue cracks.

Fig. 6.32 *Near-stall buffeting*

Flow Separation and Noise Vibrations

Boundary layer turbulence and local flow separations are other sources of vibrations (fig. 6.33). Additional problem areas for crack development are skin panels subjected to jet noise and heat (fig. 6.34).

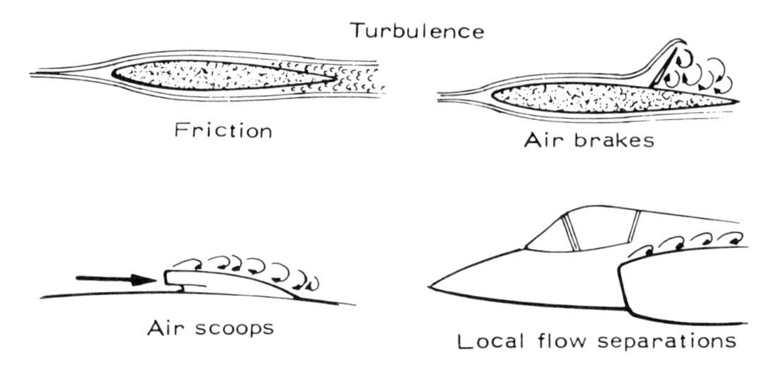

Fig. 6.33 *Sources of local vibrations*

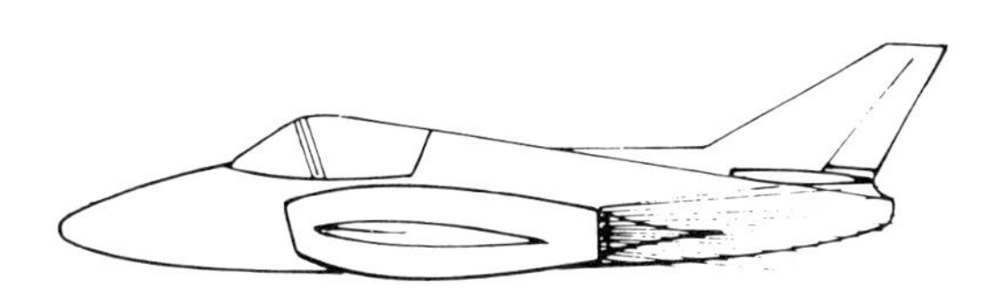

Fig. 6.34 *Heat stresses and noise vibrations*

Overloads, Corrosion, Burns and Scratches

All types of structural damage such as overload deformations, corrosion, lightning burns and tool scratches in maintenance serve as crack development points. Careless and rough handling of aircraft during overhauls may significantly reduce fatigue life (fig. 6.35). Aircraft fatigue life estimates are no longer valid if scratching is not prevented, structures are deformed, and burns and corrosion are not treated.

Fig. 6.35 *Structural damage in maintenance reduces fatigue life*

Flutter

Flutter is a divergent aeroelastic problem. If a wing during flight through turbulence bends and twists so as to increase the air load, the bending and twisting may lead to divergence and cause the wing to fail (fig. 6.36). At high speeds the oscillations may lead to wing failure in a fraction of a second.

Flutter considerations determine structural rigidity and weight distribution. If the centre of gravity of a structure (a wing) is located behind the elastic axis a positive *g* load will twist the wing in load-increasing directions. For the same reason, unbalanced control surfaces with the c.g. located behind the axis of rotation may begin to flutter. Mass balances are used to prevent this.

Flutter problems may be investigated in a wind-tunnel and during flight tests or analysed theoretically. The tests are very expensive. A flutter model of a supersonic fighter for wind-tunnel testing costs roughly the same as fifty small Volkswagens. Flight and ground testing of full-scale structures require 'flutter-guns' to start the vibrations and expensive test equipment to take measurements. For these reasons, flutter tests are usually only made for expensive military and transport aircraft. However, theoretical considerations and practical experience help the designers of light aircraft to avoid flutter in their designs.

Flutter risks can never be completely avoided. Avoidance within the flight envelope of an aircraft is achieved by designing for flutter speeds

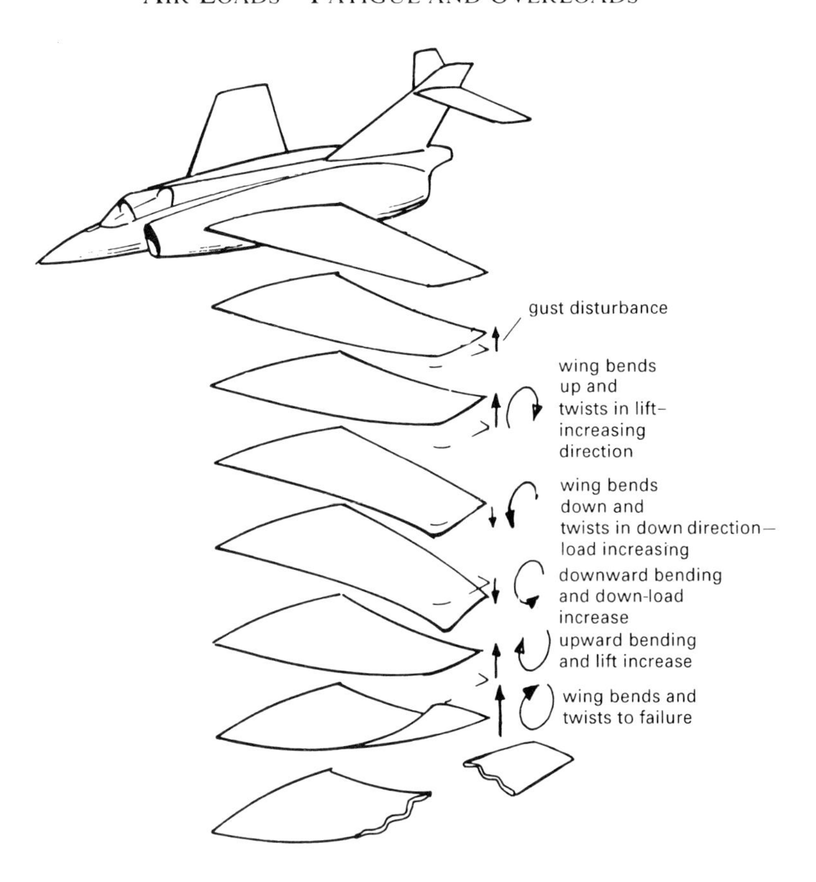

Fig. 6.36 *Wing flutter leading to failure by divergence*

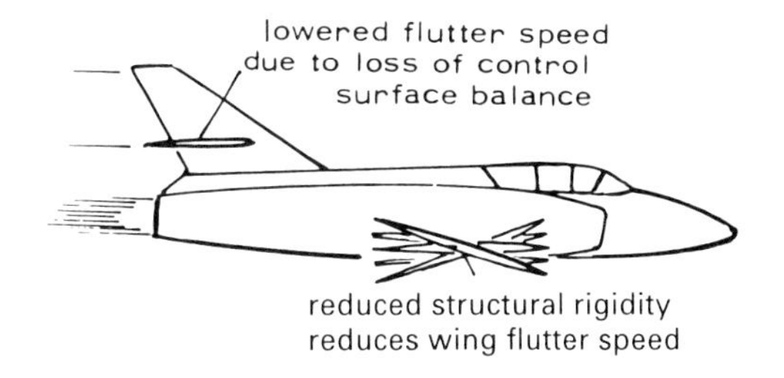

Fig. 6.37 *Partial failures or maintenance errors may have large effects on flutter speed*

considerably higher than the maximum dive speeds. All aeroplanes are flight tested for flutter up to these speeds. The risk of divergent flutter should therefore be nil.

However, this is not the case. Changes in balancing weights, control system sloppiness and reduced structural stiffness may reduce the critical flutter speed to normal operational speeds (fig. 6.37).

The following cases are examples of this.

- An elevator balance weight that was too light was fitted to a light trainer-attack aircraft during overhaul. During the test flight afterwards an 'explosive' elevator flutter developed at about Mach 0.5. The aircraft lost one elevator half but the pilot managed to land safely.

- After fatigue failure of an engine mounting strut, the engine installation of a turboprop transport developed propeller whirl flutter which tore out the engine and made the wing fail.

- A single-engine five-seat all-metal aircraft flying in icing conditions developed flutter in the empennage section. Frozen water inside the elevator had changed the mass balance. Flutter developed when the aircraft dived out of the icing conditions. The empennage section failed and fell off. The aircraft bunted and both wings failed in negative overload.

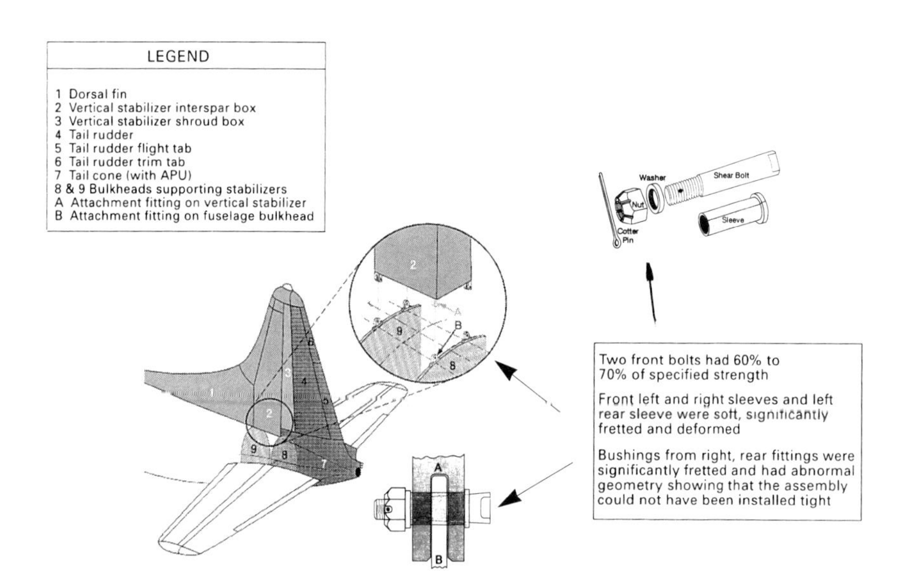

Fig. 6.38 *Bolts through A and B were too soft and not drawn tight*

- A twin-engine turboprop aircraft in cruising flight with fifty-four people on board suddenly deviated from the flight path and circled uncontrollably into the North Sea. The accident investigation showed that flutter in the empennage section had developed due to poor maintenance and bogus parts. The fin mounting bolts (fig. 6.38) were too soft and not drawn tight. This permitted the fin to move back and forth sideways in flight. The bolts and sleeves deformed and sideways vibrations developed. This started rudder vibrations which, through fuselage torsion, set up tailplane and elevator vibrations. Finally the amplitude of the oscillations became so great that both the longitudinal and directional control systems failed. The failed rudder is shown in fig. 6.39. Parts that fell off in flight were not recovered.

Fig. 6.39 *Transport aircraft fin with failed rudder*

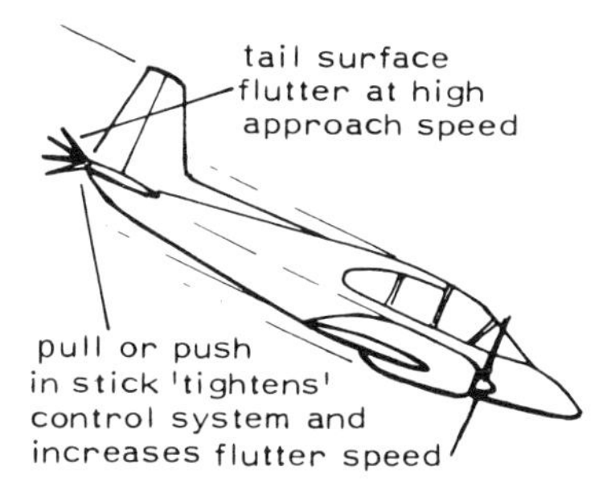

Fig. 6.40 *Elevator flutter during high-speed approach with zero stick force*

If the elevator control system of an aircraft (without boosters) is sloppy, i.e. if the elevator can be moved up and down when the pitch control is held fixed, there is considerable risk of elevator–tailplane flutter at high approach speeds with zero control force (fig. 6.40).

The flutter stops immediately if the pilot pulls stick force and takes the slack out of the control system. Flutter is dangerous. It usually causes fatal crashes. The risk of flutter is small and can be avoided by proper care and maintenance of the aircraft.

Chapter 7
Dangerous Winds

There are three types of dangerous winds affecting flight safety in addition to the turbulence discussed in chapter 6. These are:

- Winds with sharp high-speed gusts that may overload the aircraft's structure.
- Wind vortices that may roll an aircraft out of control or send it into the ground.
- Wind shears which may stall an aircraft during take-off and landing.

How Sharp Can a Gust Be?

A flight in Michigan provided the answer. It was a clear, cool day and a brilliant sun was shining from a blue sky. There was no wind, no turbulence. To the left, a black paved-road crossed spring-wet fields. I decided to follow the road and edged slowly towards it. Suddenly, with no warning my light aircraft was in a 90° bank to the right. A strong updraught from the sun-baked black road had struck my left wing. Warm air climbing through cold, still air can give very sharp high-speed updraughts.

Hot air updraughts from roads and cities seldom cause overload accidents. Even in slight winds the warm air mixes with the cold and the

(a)

(b)

Fig. 7.1 *Mountain waves and a rotor*

updraught breaks down into general turbulence. However, high-altitude mountain waves on the leeward side of tall mountain ranges are violent enough to break up an aircraft flying through (fig. 7.1a). Mountain vortices or rotors, such as the one shown in fig. 7.1b look scary enough to keep all pilots away. Flying into these would be a sure guarantee of structural failure.

However, the problem is that dangerous rotors that can not be seen can form on the leeward-side of mountains. Pilots must learn to recognise the winds that create them. Always suspect a vortex on the leeward side of a sharp mountain ridge or a steep cliffside when the wind blows across the ridge, even if the angle between the wind and the ridge is small. Crashes due to flying into leeward-side vortices with helicopters and fixed-wing aircraft are not unusual (fig. 7.5).

Thunderstorms and Tornadoes

In the central rainshaft of a thunderstorm, cold air rushes downward like water in a vertical fall (fig. 7.2). Near the ground, the falling flow spreads out in all directions. Above the outflow, warm air blows towards the outside of the shaft where it climbs to the thunderhead and, cooled down, begins a new plunge towards the ground. Extremely sharp, high-velocity shears can be met in the transitions between falling and climbing air.

All aircraft with design limit load factors below 8 g run the risk of wing failure if the penetration speed is higher than the speed for stall at the ultimate load factor (fig. 6.3). The risk of failure is proportional to the flight speed and the speed of the updraught (fig. 7.3).

There is an optimum indicated speed for storm penetration. At this speed

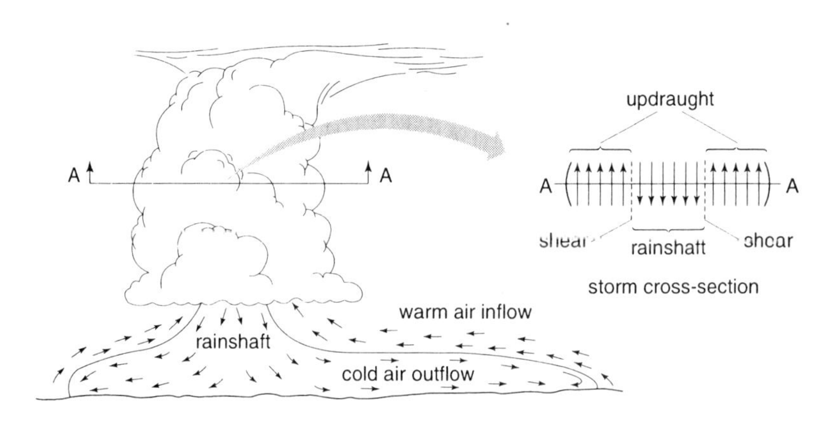

Fig. 7.2 *Vertical flow in a thunderstorm*

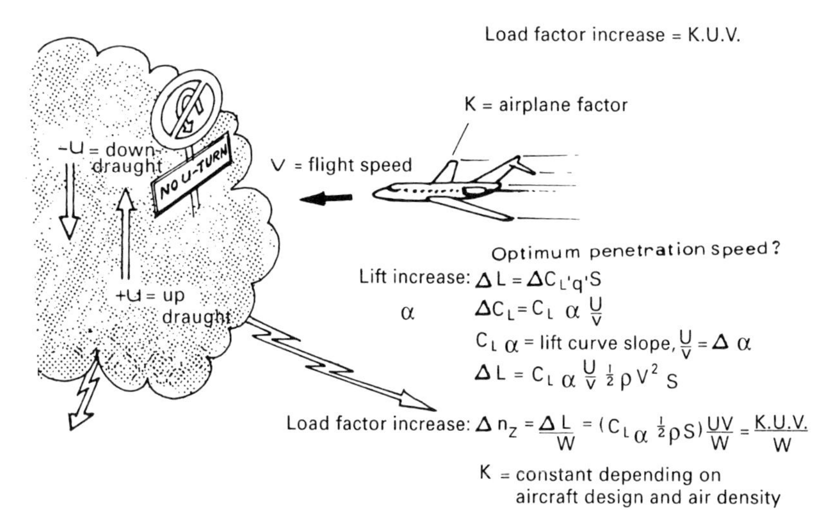

Fig. 7.3 *Thunderstorm penetration*

the aircraft stalls at a load factor just below structural failure. Penetration at higher speeds increases the risk of failure as the aircraft flies into an updraught. Penetration at lower speeds increases the risk of stall in the updraught. Pull-out from a stall in a thunderstorm is a problem since the risk of renewed stall or structural failure during a high load factor manoeuvre in gusty conditions becomes very large.

Since manoeuvring reduces the margin of stall (or failure) it is not recommended to try to get out of severe thunderstorm conditions by turning back.

Don't turn back. Increasing load factor reduces the margin to stall or wing failure!

When considering gust-penetration speeds observe that very little is known about the effects of nearly instantaneous increases of angle of attack on maximum lift. It takes time for the boundary layer to grow and spread over the wing and therefore, the 'dynamic C_{Lmax}' may be quite a bit higher than the 1 *g* value.

The speed at which structural failure at stall occurs in abrupt gusts may, therefore, be lower than the steady state stalling speed.

The lift increase (ΔL) needed to reach structural failure goes down with increasing aircraft weight and, therefore, the penetration speed must be reduced with increasing weight. However, the load factor at a given updraught intensity increases with decreasing weight. Thus, the risks of passenger injury and damage to internal equipment increases with decreasing weight.

The following accident is a typical case of wing failure during flight through thunderstorms.

A light twin with three people on board encountered heavy turbulence. The pilot contacted the closest area control and asked, 'Do you have any C.B. activity in the NN area?' The controller replied, 'No, we have no activity in the shooting range, if that is what you mean.' The pilot answered, 'Yes, that is what I mean, because I am running into some heavy turbulence.' It was a perfect case of misunderstanding between controller and pilot.

A warm front with imbedded thunderheads was located along the airway. The flight continued from thunderhead to thunderhead until the wing failed.

Data from two radar stations combined with wind information showed that the pilot had slowed down to the recommended speed, but the wing still failed. The reason for this is the high dynamic lift created by flying into sudden, sharp gusts. The dynamic lift breaks down after a very short time but it can last long enough to break short, rigid wings. Aircraft with long, elastic wings do not have the same problem because the dynamic lift collapses before the wing has been bent to its elastic limit.

The tip-tanks of the aircraft were empty. This reduced the wing failure load factor. Down aileron in the failed wing, to counter a roll disturbance, as the wing struck an updraught may also have contributed to the failure. A 4 ft outboard section of the failed wing was found 700 m away downward from the main wreckage which was concentrated in one area (fig. 7.4).

Fig. 7.4 *In-flight wing failure*

The accident shows that there is no guaranteed safe way to penetrate a thunderstorm. A wing may fail even if the flight speed is reduced to the recommended penetration speed. Since the probability of running into wing-failure conditions during a flight through a C.B. cloud is small, 'experience' may tell pilots that safe penetrations can be made. This is wrong! Check weather before departure. Avoid C.B.s *en route*!

An airline pilot landing at an airport close to the accident site said, 'This was the worst turbulence I have experienced. I could hardly read the instruments'. This was reported to the local meteorological office. No action was taken since no more traffic was expected at the airport that evening. In the meantime, the flight above continued through the thunderstorms.

The accident was a typical case of lack of weather briefing and risk awareness by the pilot, lack of severe-weather reporting to A.T.C. and lack of risk knowledge by the controller. Flying a fairly short distance south or north of the front would have prevented the accident.

A short time after this accident, a twin-engine jet transport flew into tornado-like winds above Holland. The accident investigation showed that the wing, which was stronger than required, had failed in overload.

Even in this case, information about the dangerous weather was available but was not reported to the pilots. Lack of risk awareness causes many easily avoidable accidents.

Leeward-side Mountain Vortices

The aircraft in fig. 7.5 took off from a long, narrow lake in what appeared to be a headwind. What the pilot failed to observe was that the wind at altitude blew across the valley and that a strong leeward-side vortex probably rotated near the steep mountain side to the left of the take-off path. After take-off, the pilot climbed and turned to the left. The aircraft was caught by the vortex downdraught and thrown into the ground.

A Twin Otter took off from an island airstrip built along a steep mountain ridge. During the take-off climb it was thrown upwards, to the right and into the ocean. Water board tests later showed that strong leeward-side vortices formed over the runway in crosswinds over the ridge (fig. 7.6). People living near the airport told investigators that they had seen seagulls thrown into the ground in crosswinds.

Ground-induced turbulence in heavy winds may also be sufficiently strong to overload aircraft structures. A light transport flying at low altitude off the coast of northern Norway in a gale dived into the ocean after an empennage failure.

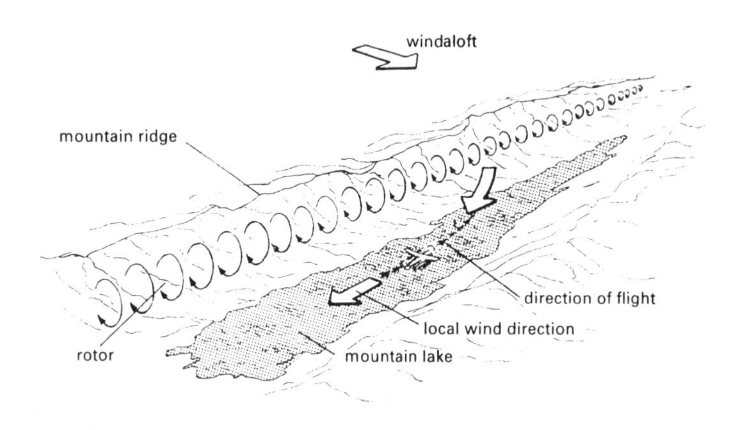

Fig. 7.5 *Flight into leeward-side vortex*

Fig. 7.6 *Leeward-side vortex along runway*

Wake Turbulence

According to the National Transportation Safety Board (N.T.S.B., USA) fifty-seven accidents or serious incidents happened between 1983 and 1993 when aircraft during landings and take-offs, flew into the wing-tip vortices trailing from aircraft in front. The wind velocities in these horizontal tornados may be ten times as large as the limiting gust velocity for civil aircraft. Aircraft flying across the vortices may be subjected to critical up or down loads and aircraft flying into the vortices on a more parallel course may be rolled on their backs (fig. 7.7).

The largest vortices are created by slow-moving aircraft, i.e. aircraft in the landing phase. The vortices move downwards behind the aircraft with a speed of roughly 2 m/sec (approximately 350 ft/min) and as they reach the ground they move away from each other with a speed of approximately 5 knots (roughly 2.5m/sec).

This means that a vortex in a crosswind of 5 knots may remain on the

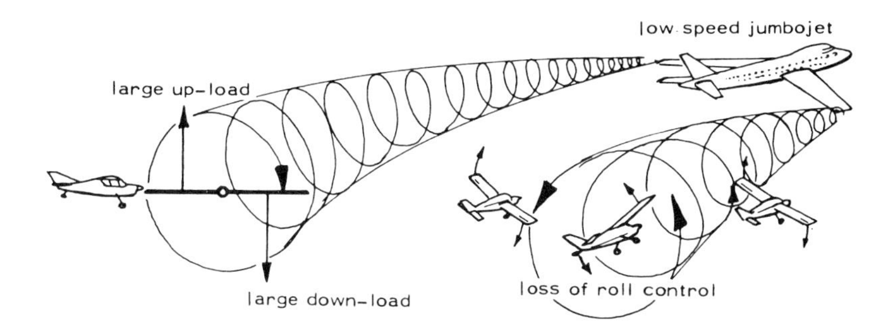

Fig. 7.7 *Effects of trailing vortices*

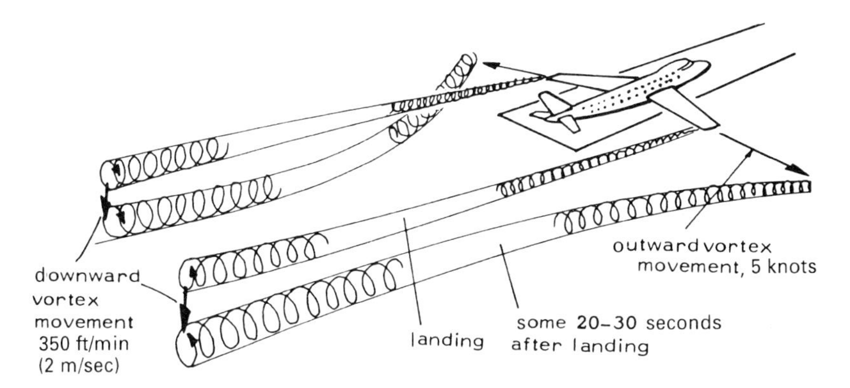

Fig. 7.8 *Vortex movement behind a landing aircraft*

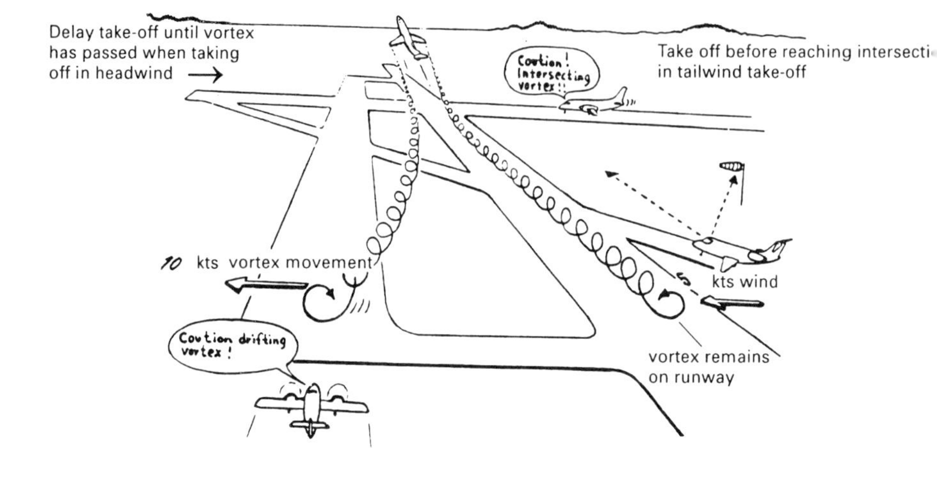

Fig. 7.9 *Risks of vortex interference*

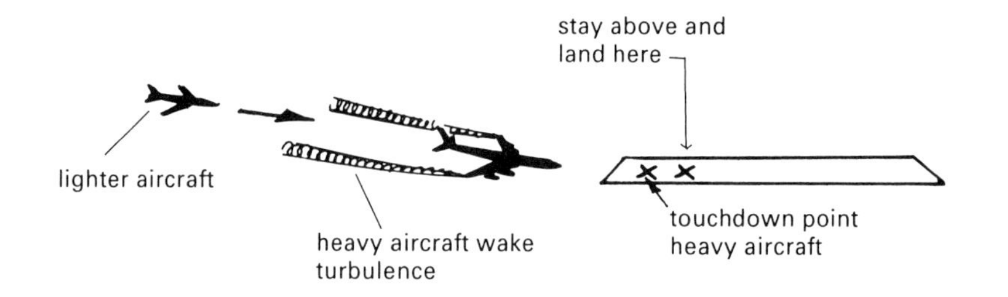

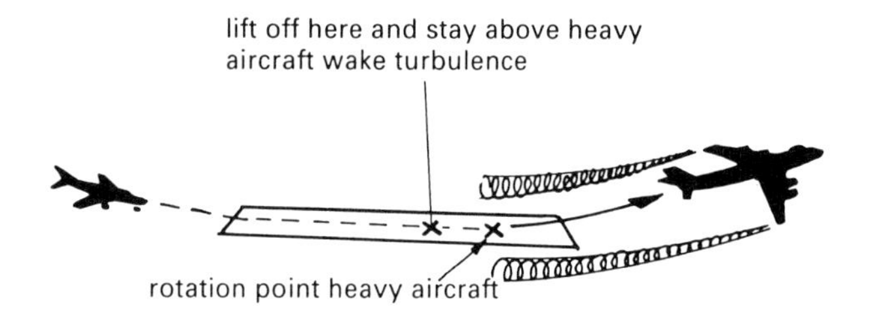

Fig. 7.10 *Avoiding vortex interference during take-offs and landings*

runway several minutes since it contains enough energy to rotate for a long time in steady, turbulence-free winds. Vortices rolling along the ground may create problems for aircraft landing on parallel runways. When intersecting runways are used for take-offs and landings an aircraft may run into vortex disturbances from aircraft crossing its path (fig. 7.9). The risks are highest on calm days when there are the steady, low-turbulence winds.

In order to minimise the risk of dangerous vortex disturbances during landing, a minimum time space of 2 to 3 minutes (approximately 4 to 6 nautical miles) between aircraft on approach has been set for landing behind heavy transports. Recent accidents have shown that this time is not always sufficient.

Smaller aircraft landing behind heavy aircraft should stay above the latter's approach path and touch down beyond the touchdown point of the heavy aircraft in order to stay away from its large vortices (fig. 7.10a). Smaller aircraft taking off behind heavy aircraft should lift off before reaching the latter's rotation point, where the generation of large vortices begin, and stay above the heavy aircraft's climb-out path to avoid interference (fig. 7.10b).

Helicopter vortices are large compared to the size of the generating aircraft. Small aircraft must, therefore, be warned against helicopters operating near active runways. In one case, a helicopter crossed a runway a short time before a light aircraft took off. When it reached the path traced by the helicopter it rolled on to its back and crashed.

Other cases of vortex interference include the following:

- A DC-9 making touch-and-go landings crashed when touching down on a runway where a 747 had just landed.
- A medium-heavy jet transport rolled on its back, but recovered, when, during its take-off climb, it flew into the vortices trailing from a heavy jet above and ahead of it.

When two low aspect ratio aircraft make a formation approach, there may be considerable risks for vortex interference. This is especially the case if the leading aircraft makes a wave-off and creates a large vortex at low level that may move sideways as it hits the ground and roll right in the path of the second aircraft. The problem may be exacerbated in crosswind landings with the second aircraft on the downwind side.

Wing-tip vortices are very large compared to the size of the landing aircraft as illustrated by the size of a Boeing 747 vortex in fig. 7.12a and the size of the vortices of a fairly small delta wing aircraft landing on a snow-covered runway shown in fig. 7.12b. This illustrates the importance of staying away from trailing vortices.

In formation flying, aircraft flying into the vortices of the lead aircraft

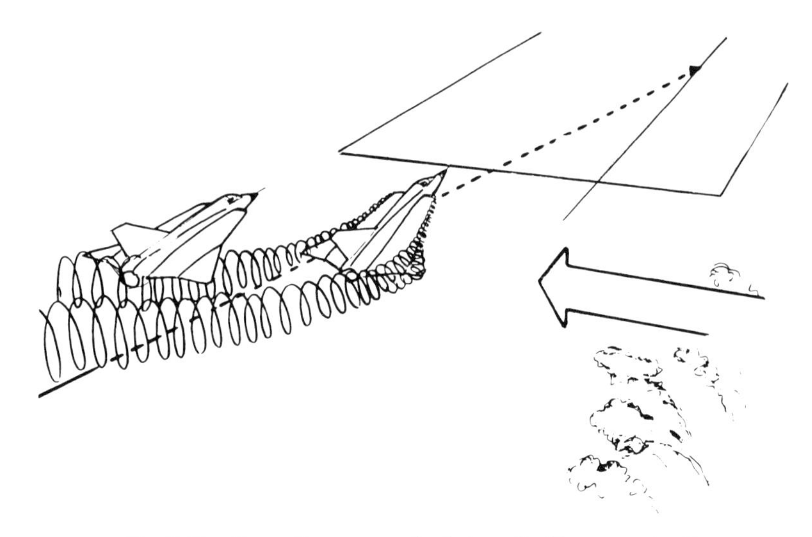

Fig. 7.11 *Formation approach of aircraft of low aspect ratio*

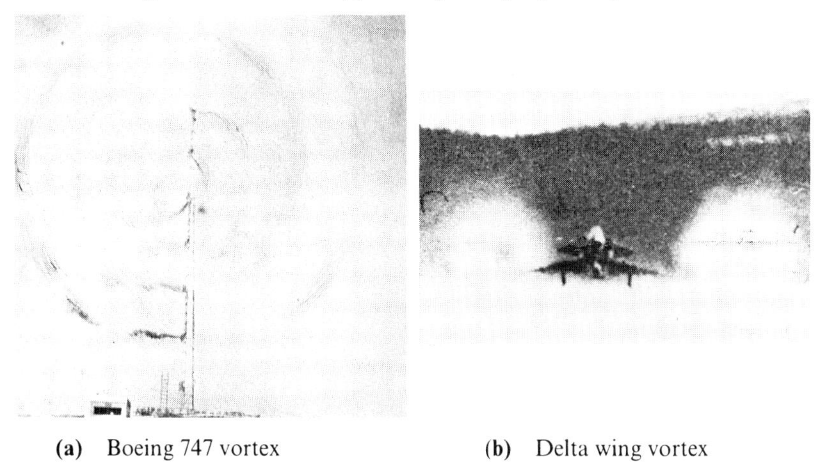

(a) Boeing 747 vortex **(b)** Delta wing vortex

Fig. 7.12 *Vortex diameters*

have rolled into the formation and collided with other aircraft in the group. Have respect for wing vortices!

Wind Changes and Wind Shears

When winds change rapidly from headwinds to tailwinds it may be in possible for aircraft to accelerate and maintain safe stall margins during take-offs and landings.

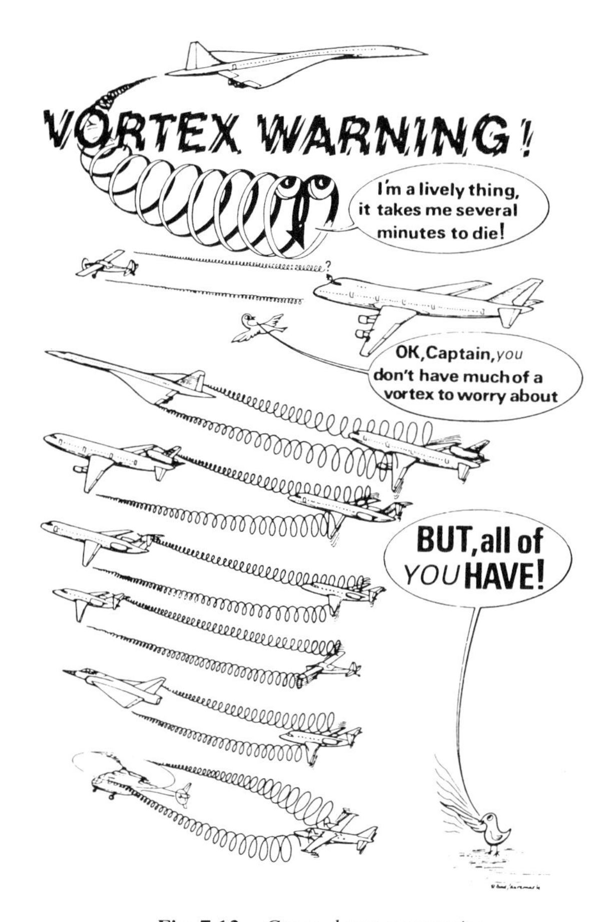

Fig. 7.13 *General vortex warning*

The aircraft in fig. 7.14 stalled immediately after lift-off when the headwind changed to a tailwind. It took off from a grass strip in a shallow valley located between two parallel ridges. Aloft, the wind blew across the valley but at ground level, as it dived into the valley, it blew across the middle of the runway, then split and turned on both sides toward the runway ends (similar to the wind illustrated in fig. 7.15). Thus, the pilot had a headwind at the start of the take-off run, a crosswind at the middle and a tailwind that, by coincidence, gusted up to stall the aircraft at lift-off.

The airstrip had only one windsock at the clubhouse. Under such conditions, several wind indicators are needed. Many airports, even large international ones, are plagued by unpleasant wind changes in the approach, landing and take-off zones. The wind variations must be reported to the pilots. Unfortunately, average winds over several minutes

Fig. 7.14 *Stall on lift-off due to a sudden tailwind*

are reported. During these minutes, aircraft may crash due to unreported wind changes.

The most critical wind changes are probably obtained in the wind shears below thunderstorms. A jet transport taking off below a C.B. cloud at an airport had a 10-knot headwind in the outflow from the storm at the beginning of the take-off run. At lift-off near the downpour from the rainshaft, the wind had changed to 40 knots. As the aircraft climbed it entered the downdraught where the wind rapidly changed to a 70-knot tailwind as the flight continued. As a result, the aircraft stalled and mushed back into the ground (fig. 7.16).

During approach to an international airport below a thunderstorm, a jet transport first flew into the updraught towards the outside of the rainshaft. As a result, the aircraft 'ballooned' above the instrument landing system (I.L.S.) path. As thrust was reduced to correct for this, the aircraft lost altitude in the downburst from the rainshaft and finally lost flight speed in the tailwind from the outflow. With insufficient air speed the aircraft sank into the ground near the runway threshold (fig. 7.17).

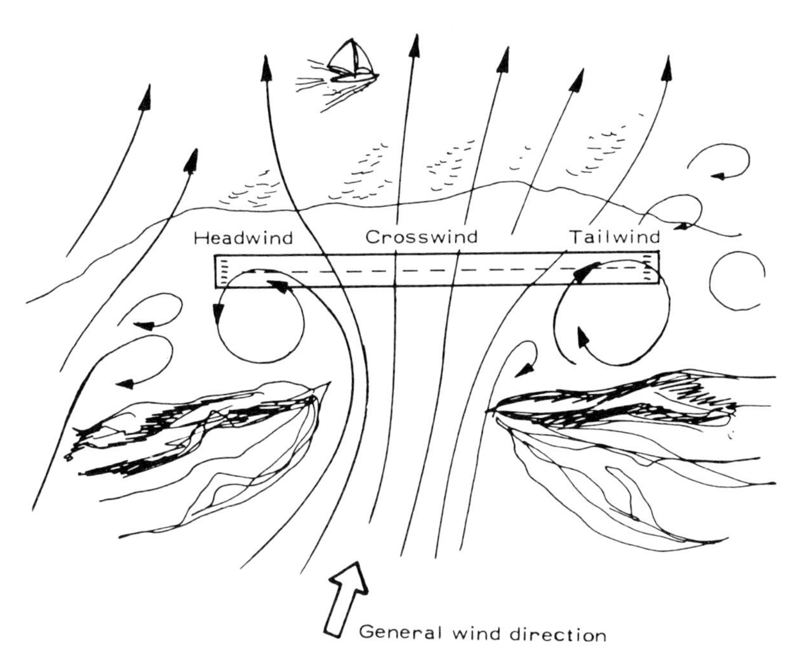

Fig. 7.15 *Local wind changes due to topography*

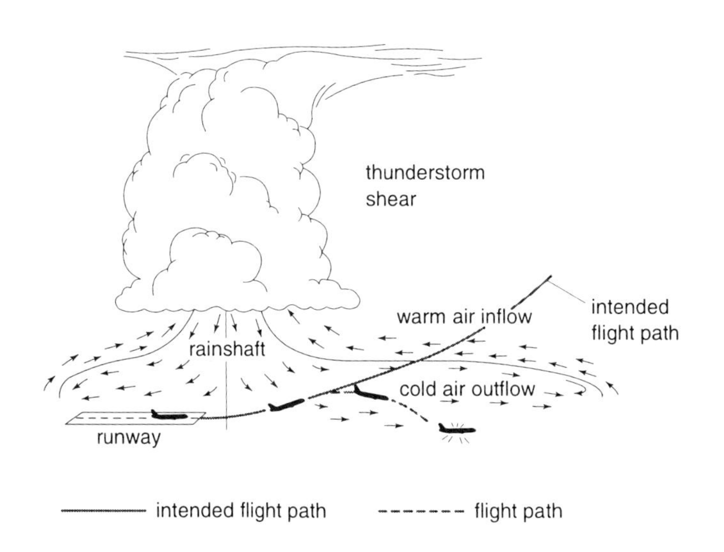

Fig. 7.16 *Tailwind shear and stall during take-off below a thunderstorm*

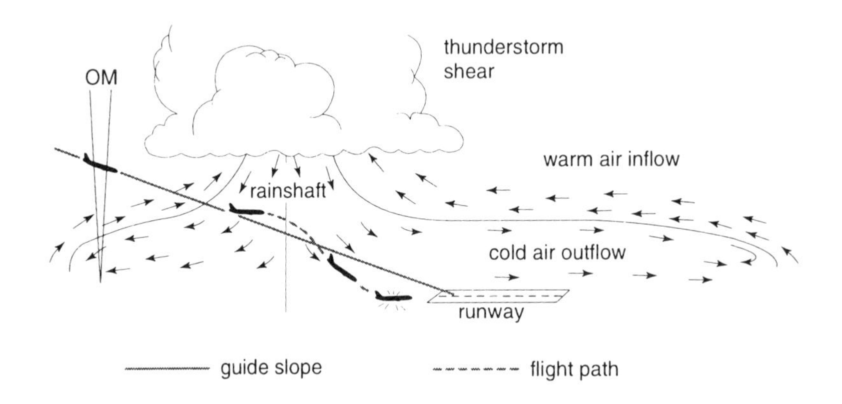

Fig. 7.17 *Approach accident due to wind shear below a thunderstorm*

One problem with thunderstorm shear detection is that a C.B. cloud which may be difficult to spot in general cloudiness with no lightning may build up fast and move rapidly. The following incident illustrates this.

The Boeing 720 (in the runway overrun zone in fig. 7.18) was on approach. A 3-knot tailwind was reported and in the approach zone there were scattered summer clouds. However, above the threshold a dark C.B. cloud was growing in size. The met-assistant studied the phenomenon but reported nothing since no lightning was seen. This was the rule. Suddenly, the tailwind increased to 25 knots. Conflicting traffic on another runway made it impossible for A.T.C. to ask the 720 to go round. Then, the 25-knot tailwind dropped to 11 knots and the flight was cleared to land.

Fig. 7.18 *Wind shear overrun*

The landing was made in tropical rain. The air speed at touchdown was critically low but the ground speed was high so the aircraft aquaplaned

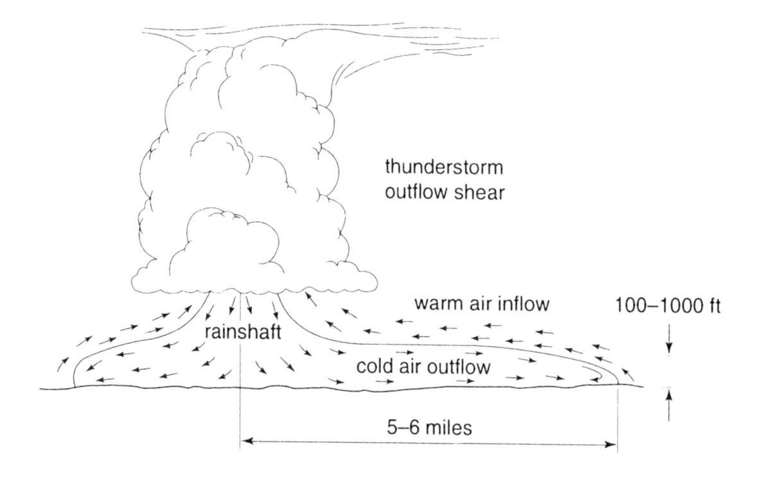

Fig. 7.19 *Thunderstorm outflow*

down the runway with no possibility of stopping before it reached the end. It was a clear case of not understanding the wind shear risks.

It took a long time and many accidents before aeronautical engineers and accident investigators finally realised that the winds below a C.B. cloud can change faster than aircraft can accelerate and that the influence area may stretch miles from the storm centre (fig. 7.19).

Detection of microbursts is a problem in wind shear avoidance. Airports may be surrounded by a number of small thunderstorms, microbursts,

Fig. 7.20 *Dry-air microburst*

which move around. They are difficult to see and follow on radar scopes as they drift in the area. In dry climates the humidity in the microburst chute evaporates but the downburst still strikes the ground as shown by the dust ring at the rim of the outflow (fig. 7.20).

Wintertime inversions may create critical wind shears. Supercooled air close to the ground may drift in one direction while a warmer wind above blows in the opposite direction (fig. 7.21). High-speed cold air from inland valleys may spill over the crests of mountain ridges, accelerate down the slopes and flow over airports at the same time as mild ocean winds blow in the opposite direction. Aircraft taking off in a cold headwind may climb into a sudden, warm tailwind and lose indicated speed both because of the shear and because of the temperature increase. The following accident is typical.

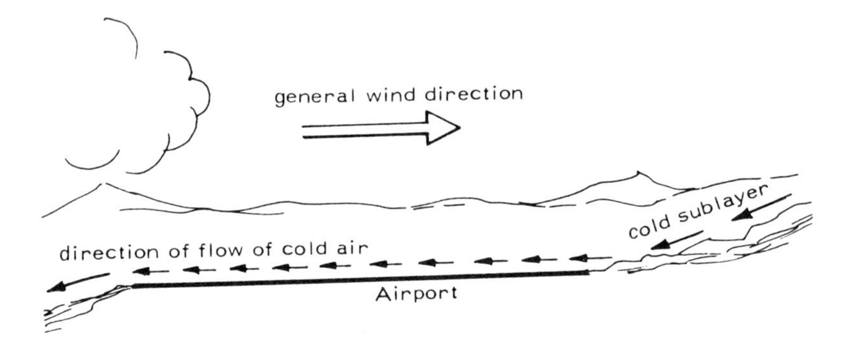

Fig. 7.21 *Inversion wind shear*

It was a cold, clear winter evening with large radiated heat losses cooling the air closest to the ground. A four-engine jet transport accelerated for take-off in a 3-knot headwind on a three-engine ferry flight with only the crew on board. As the aircraft lifted off it flew through light ice crystals blown up by another aircraft. The captain lost external references. The aircraft yawed a few degrees which increased the stalling speed by 5 knots. At the same time, the aircraft climbed into an unreported (unknown) 10-knot tailwind which was 10°C warmer than the ground-level temperature. The captain now had to pull back on the yoke to maintain lift. As a result, the angle of attack increased to near stall where local flow separations on the wing combined with high induced drag produced a total drag that was higher than the maximum thrust and the aircraft descended into the forest beyond the end of the runway.

Thunderstorms, terrain and inversions can create all types of wind shear which may cause accidents unless wind shear warnings are given and heeded by the pilots. The following are the risks.

- Crosswind shears near runway touchdown zones may throw landing aircraft off course and make touchdown on the runways impossible.
- Headwind to tailwind shears may stall an aircraft or make it land short. However, since higher thrust is required in a headwind approach along a fixed glide path than in a tailwind the aircraft will enter the tailwind with too much thrust. If the shear is mild and too much thrust is added to regain approach speed, the aircraft may accelerate and land far down the runway at too high a speed. On slippery runways the risk of overrun may be large.
- Tailwind to headwind shear may appear to be safe since flight speed is gained and the margin to stall increases. However, if the tendency of the aircraft to balloon and land far down the runway is corrected by too large a thrust reduction, the aircraft may lose too much speed since more thrust is required to fly along a fixed glide slope in a headwind than in a tailwind. As a result, the aircraft may develop a high sink rate and land short of the runway.

Aborted landings during tailwind to headwind approach shears create a special problem. This is illustrated by the following accident.

The twin-engine jet had flown into a headwind from a tailwind approach and had slowed down to normal approach speed when the crew decided to discontinue the approach due to heavy rain at the airport. The crew added power, pulled the aircraft into a climb and turned right to avoid the downpour. When they did this, the aircraft stalled, lost altitude and struck the ground in a 5° nose-high attitude. The aircraft had re-entered the tailwind in a climbing turn at increased angle of attack due to pull-up and turn. At the same time, the heavy rain had probably increased the stalling speed of the aircraft. Stall could not be prevented. It is good practice to discontinue approaches into severe weather. However, abrupt pull-ups from headwinds to tailwinds are dangerous.

Wind shear problems are also discussed in chapter 9.

Chapter 8
Engine Problems

Engines, like aircraft structures, are designed for estimated operational environments, i.e. for certain combinations of maximum power, cruise power, idle power, heating, cooling, wear etc. Repeated use of the engine outside the design environment increases the risk of premature engine failure.

Reciprocating Engines

The power and safe life of a reciprocating engine are affected by a number of factors, such as:

- engine wear which increases internal friction
- particle ingestion (dust, sand, volcanic ash) which may create excessive wear
- faulty ignition adjustment, spark-plug deposits and too low an octane fuel all of which may cause faulty detonations and engine overstresses
- faulty fuel–air mixture; too rich a mixture leaves soot deposits on spark-plugs (and detonations) and reduces range; too lean a mixture may overheat the engine and cause detonations
- high frequency of engine heating and cooling which results in early development of fatigue failures
- ice in air intake or carburettor ice choking the engine
- frequent flights at maximum power
- frequently exceeding maximum allowable r.p.m.
- engine starts without preheating on cold winter days.

The following accidents show what may happen. The light aircraft had completed a large number of glider towings during the summer. Each flight consisted of a maximum-power take-off, a high-power climb and, after glider release, an idle-power dive towards the airport to drop the towline. After the drop, the pilot made a high-power pull-up and turn followed by an idle-power precision landing with an occasional power burst to correct for wind disturbances. The engine stopped due to fatigue failure of a piston rod during a pull-up after a line drop. Engine lives may become very short with excessive use of high power combined with frequent engine cooling in idle-power descents.

Fig. 8.1 *Stall after engine failure during take-off climb*

The aircraft shown in fig. 8.1 had been parked outside for a couple of days in cold winter weather. The pilot scraped ice off the wings but did not bother to preheat the engine (the oil) before attempting engine start. After all, the sun was up and the temperature in the cabin was not too unpleasant. Engine start proved to be difficult but after a few attempts the engine quit coughing and began to run. During climb-out a piston failed and the engine stopped. The pilot was lucky. The aircraft stalled just before touchdown during his return glide to the airport. His face was banged blue, yellow and green against the glareshield. Stalling a few metres higher up could have been fatal. Frozen oil may overload an engine during engine start.

A light, piston-powered helicopter was used for counting moose. This is done in wintertime when the animals are easy to spot against the snow. However, since the sex has to be determined (this is in Sweden) and not all bulls have winter horns, the pilots must fly very low and look at the animals' rear ends. This requires a lot of control juggling, high power and high risk of engine overspeeds. The helicopter in fig. 8.2 crashed due to an engine fatigue failure 42 flight hours after an engine overhaul. Continued flight at near maximum power combined with high rates of power changes and repeated engine overspeed can wear out an engine in a very short time.

Fig. 8.2 *Fatigue failure of engine*

The Jet Engine

The modern turbine engine is a top performer. Its fan, compressor and turbine blades have been shaped with minute care. With the aid of advanced theories and computers the engine aerodynamicist has squeezed the last drop of performance from the blades.

These engines are sensitive to wear and tear. After a couple of months in service their fuel consumption increases by a couple of per cent compared to the factory-new consumption.

Surface irregularities, including contamination, increase individual blade drag and reduce the stall angles. As a result, maximum engine thrust decreases and fuel consumption increases. Compressor stalls caused by blade surface roughness or by disturbed air intake flow may cause internal, structural engine damage resulting in engine failure. The engines are sensitive (fig. 8.3).

If you fly through an abrasive environment, such as a cloud from a volcano eruption, with these engines there is a high risk of engine flame-out and excessive blade wear.

Helicopters operating in desert areas must be equipped with particle separators in the air intakes in order to avoid excessive compressor wear. Hard fertiliser pellets used in forest fertilisation may grind compressor blades as sharp as razor blades in a short time. When this happens, the payload the helicopter can lift decreases to zero.

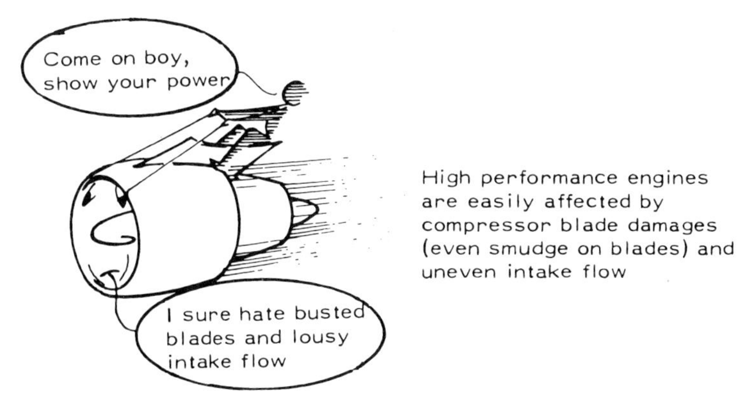

Fig. 8.3 *Sensitive jet engine*

Salt deposits on blades decrease power and increase fuel consumption. Compressors may have to be washed once a day. Operation in insect-infested environments (mosquitoes, flies, etc.) may also require frequent engine cleaning. The following case is an example of contamination.

An air force having problems with engine corrosion due to a salty atmosphere decided to throw a can of oil (without the can) in the air intake when the engines were spooling down after the last flight of the day. Corrosion was prevented but the thin oil film on the blades gave compressor stalls at high angle of attack combat-training flights. Blades must be kept clean and smooth.

Wing-mounted engines, hanging fairly close to the ground, may suck in material from the ramp, taxiway and runway surfaces (fig. 8.4). It is, therefore, rather obvious that these surfaces must be kept clean. Far too often there are small stones, gravel, etc. on runways and taxiways, and paper and plastic bags blowing around the airports. The picture in fig. 8.5 was taken at the ramp of a major international airport. If paper is sucked into an engine it may stall the compressor. A compressor stall sends shock loads through the engine and distorts the airflow. Distorted airflow, again, may create hot

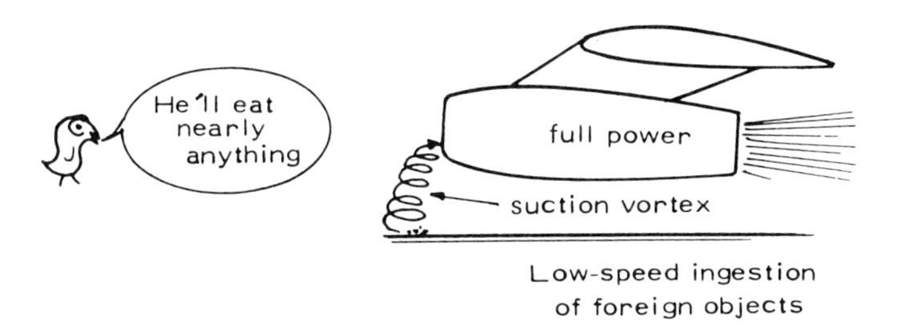

Fig. 8.4 *Ingestion of foreign objects*

Fig. 8.5 *Litter on the ramp*

spots that may damage the engine. Airport surfaces must be kept clean!

The use of thrust reversers at low speeds may result in reingestion of the reverser air by the air intakes. The result is usually very poor intake flow caused by turbulence and flow separations at the intake lips. This may also result in compressor stall, distorted flow, hot spots and large compressor blade loads which reduce engine life. Dust and debris thrown up by the

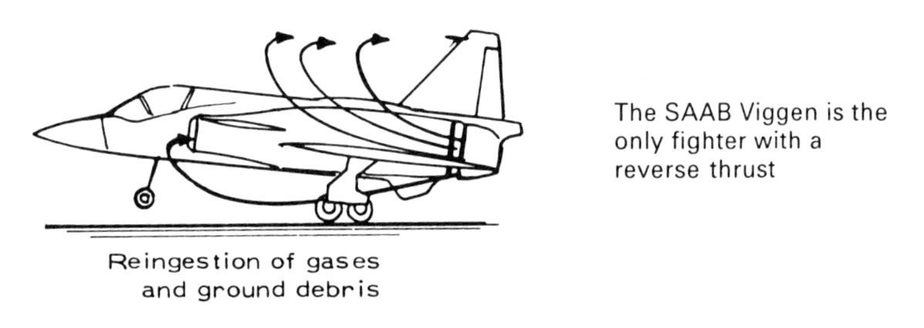

Fig. 8.6 *Reingestion when reversing at low speeds*

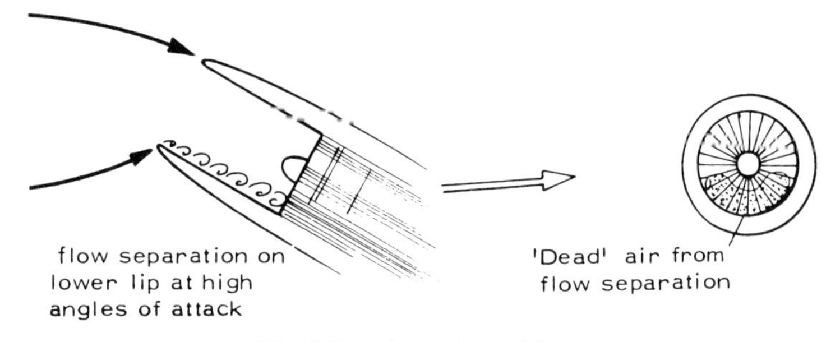

Fig. 8.7 *Air intake problems*

reversers damage the engines (fig. 8.6). Jet transports with wing-mounted engines are most seriously affected.

Flow separation at the air intake lips may slow the intake flow over part of the compressor face (fig. 8.7). When a blade rotates through the disturbed-flow zone the reduced speed of the intake air increases the blade angle of attack. As a result, the flow disturbance has two effects:

- risk of compressor stall, hot spots and shock loads in the engine
- continuous, high-frequency variation of the blade loads may result in blade fatigue failures (especially if the blades are scratched).

The problem is greatest at high angles of attack. It should be solved for all aircraft within the normal operational envelope. However, transport aircraft have run into compressor stalls during crosswind ground-runs (fig. 8.8). This has resulted in aborted take-offs and overrun accidents.

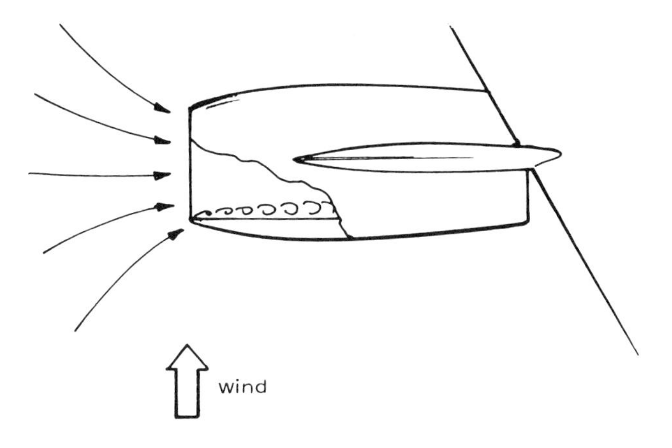

Fig. 8.8 *Intake flow separation in a crosswind*

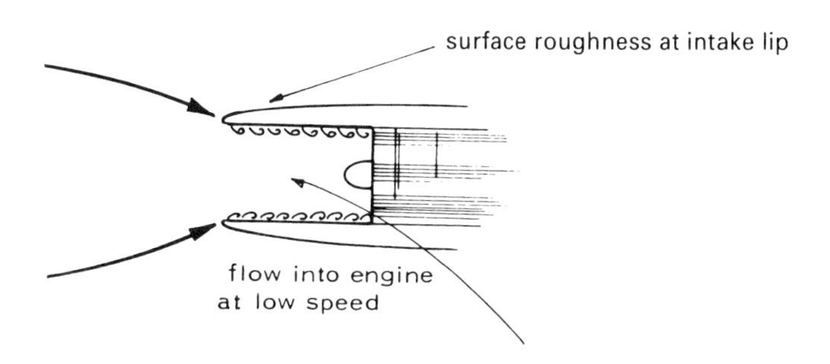

Fig. 8.9 *Intake-lip flow separation during take-off run*

The intake flow separation problem is exacerbated by intake lip wear, damage (dents) and contamination. The problem is greatest at maximum power during the early part of the take-off run while the diameter of the flow sucked into the engine is larger than the intake diameter (fig. 8.9) and the flow separation at the lip is ingested by the engine.

The flow disturbances caused by ice formation on the lips are very large. It is therefore very important to maintain ice-free lips (engine heat on) during take-offs and flights in icing conditions.

Switching on the engine anti-ice system after ice has formed on the intake lips may have disastrous effects (fig. 8.10). When the ice begins to melt and is sucked into the engines it may knock out compressors or cause flame-outs. Snow ingested by helicopters in flight is a common cause of flame-out (which may be cured by means of continuous ignition).

Slush sliding off fuselages into fuselage-mounted engines during take-off rotation, and clear ice ingested by the engines when the ice breaks loose from the wing-root sections as the wings bend during lift-off have caused several accidents (*see also* chapter 10).

In-flight collisions with large hailstones may destroy an engine compressor (fig. 8.11) and so may high-power flights into tropical rain storms.

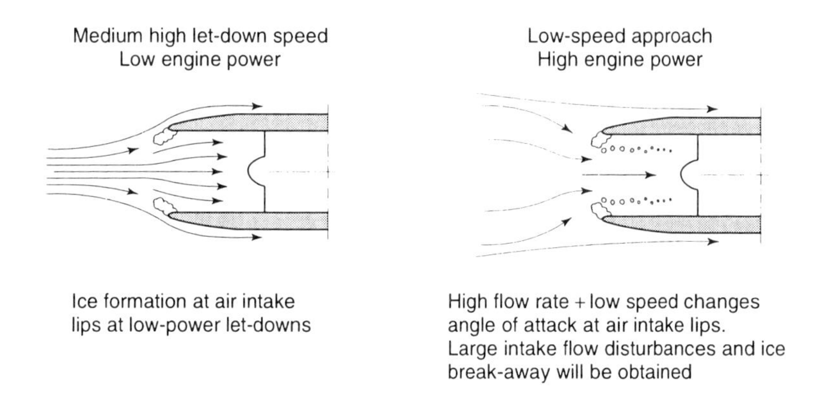

Fig. 8.10 *Engine failure due to ice ingestion*

BIRDS! Birds need places to eat and rest and if your airport looks inviting they will come there (fig. 8.12).

If your airport is located close to the seashore, and the city dump on the inland side of the airport is used by the local meat factory to dump their waste, as was the case at the airport shown in fig. 8.13 you are in for some striking problems with birds.

Fig. 8.11 *Hail damage to engine compressor*

Fig. 8.12 *Birds on the move*

Fig. 8.13 *Birds at the airport*

The air force and the domestic airline had complained for ten years about the bird-strike risks without getting any response from the local politicians. Fortunately fifteen of the 'city fathers' were on board an F28 on the day of the 'big strike'. Shortly after lift-off a number of fat seagulls slammed into the wings and one of the engines. During the subsequent emergency landing a sickening smell of burnt feathers filled the passenger cabin. The following day, the city council unanimously voted to move the dump to a safer place. The bird-strike problem was solved.

A medium-size bird can cause considerable structural damage and knock out an engine if it impacts at a critical point.

Fig. 8.14 *The result of bird-strikes in an engine*

Keeping the airport clean means keeping birds away from runways. A bird-strike in an engine may cause disaster. The aircraft shown burning in fig. 8.14 was the first to take off in the morning. Nobody had bothered to drive along the runway beforehand to chase away the birds. As a result, the right engine ingested a number of seagulls, and caught fire. This happened before V_1. The pilot managed to stop at the runway end. The last stewardess jumped out of the aircraft just before the fuel exploded. There were only 100 people on board. Evacuation of a fully loaded aircraft in time to prevent fatalities would not have been possible even if the fire-tenders had been on the spot at the same time as the aircraft stopped.

Numerous incidents and accidents show the importance of being earnest when considering engine failure risks and risk avoidance.

CHAPTER 9
PERFORMANCE

Aircraft performance is determined by the following factors:

- aircraft weight
- lift and stall for various flight configurations (clean aircraft, effects of high-lift devices etc.)
- aircraft drag as a function of lift for various flight configurations (clean aircraft, effects of high-lift devices, effects of landing-gear extension etc.)
- engine power/thrust; fuel consumption and fuel volume.

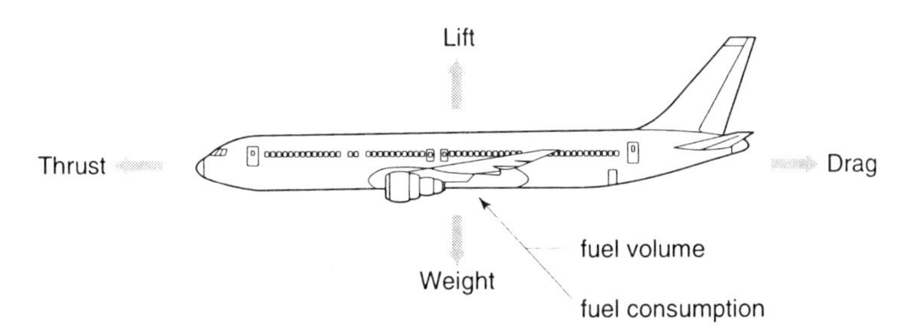

Fig. 9.1 *Basic factors affecting performance*

The performance data in the flight manuals are based on theoretical calculations, wind-tunnel tests and flight tests. Typical design data flow and adjustments during the development of an aircraft are shown in fig. 9.2.

The design goals are not always met when the first series aircraft are delivered. Aircraft drag and stall are difficult to determine theoretically and in wind-tunnel tests. Since no manufacturer overestimates drag and underestimates lift when presenting a new project they are usually stuck with drag-reduction and lift-improvement programmes while delivering the first series aircraft.

The performance of an installed engine is usually not as good as the

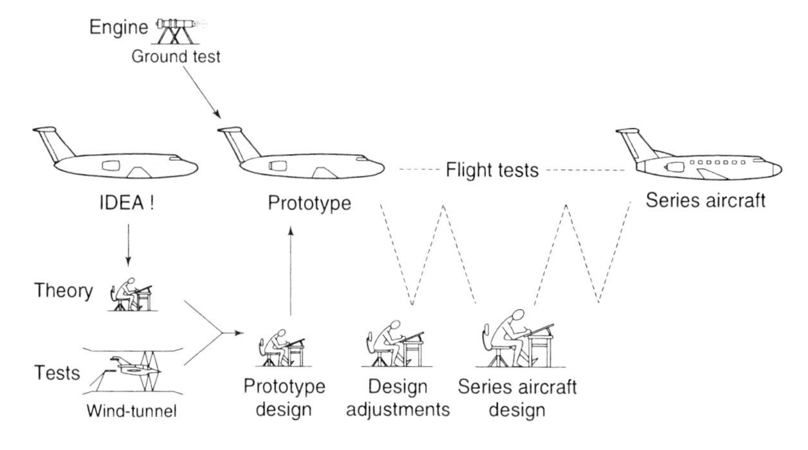

Fig. 9.2 *Design progress*

performance obtained in test rigs due to installation losses such as air intake and exit losses. When flight test performance is poorer than estimated it is difficult to determine if this is a result of underestimated aircraft drag or overestimated engine performance. This may lead to animated discussions between airframe and engine manufacturers (fig. 9.3).

One problem which must be addressed when considering handbook performance data is that the flight tests used for data verification are made by highly trained test crews flying new aircraft.

In real operations, performance degradation due to engine and airframe wear must be expected. Also, the take-off and landing

Fig. 9.3 *Airframe or engine problem?*

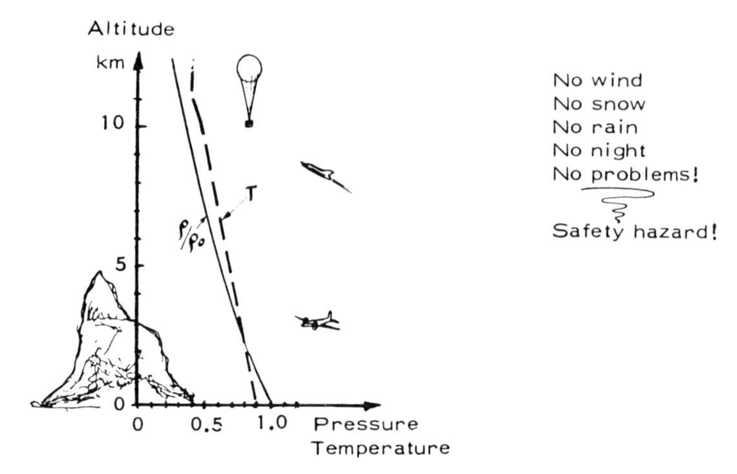

Fig. 9.4 *The standard atmosphere*

performance demonstrated by test pilots operating in excellent conditions may not be easy to repeat by less well trained pilots in poor weather. This makes it necessary to maintain safe margins to the optimum data presented in the manuals.

Also, all performance data are made for standard atmospheric conditions. However, the standard atmosphere is a dead model of the real one. It does not contain the meteorological problems, the contamination problems, the mountains and the birds of the real atmosphere. Even if corrections are made for pressure and temperature changes, a number of problems remain which the pilot must take into consideration when planning a flight.

Air Data

Correct air data registration is essential for safe flight. Any factor affecting the pressure and angle of attack readings may have catastrophic effects.

Indicated air speed is obtained by measuring the difference between the total pressure and the atmospheric pressure. This gives the dynamic pressure which is calibrated to give air speed (fig. 9.5).

Indicated and true air speeds are equal only at sea level and standard day temperature and pressure. True air speed becomes greater than indicated when temperature increases and pressure decreases. The change of true air speed at constant indicated speed with increasing altitude in a standard atmosphere is shown in fig. 9.6.

The total-pressure probe on subsonic aircraft is usually mounted on the wing leading edges (small aircraft) or on the side of the fuselage nose (trans-

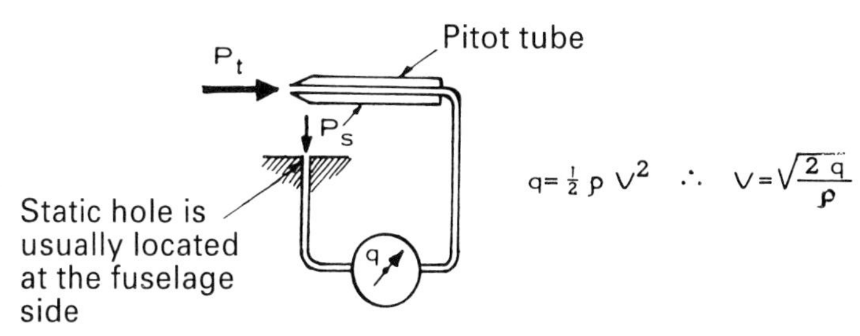

Fig. 9.5 *Air speed measurement*

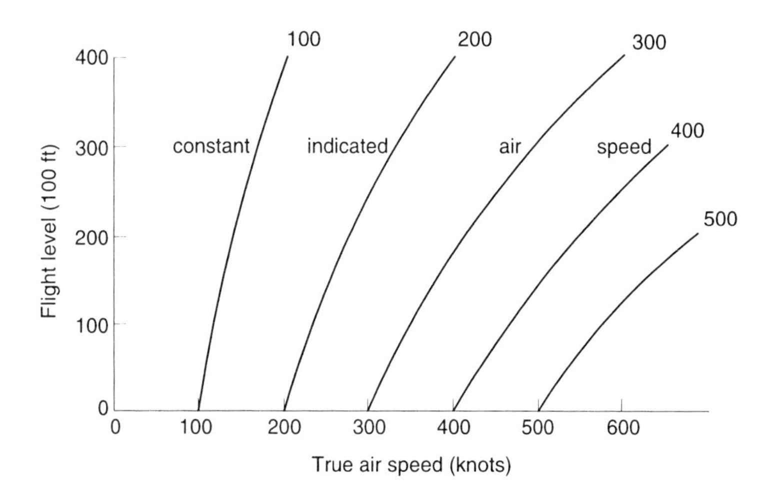

Fig. 9.6 *True air speed increase with altitude*

port aircraft). The static-pressure port is usually located somewhere on the fuselage side where the local flow speed equals the flight speed and the static pressure equals atmospheric (fig. 9.7).

The angle of attack vane near the fuselage nose supplies angle of attack data to the stall warning system. On light aircraft the stall warning device is usually located at the wing leading edge. A small metal tongue protrudes through the wing nose (fig. 9.8). At a certain angle of attack when the flow stagnation point at the profile nose moves below the tongue, the airflow flips the tongue upward and triggers the stall warning device.

Since the altimeter is calibrated to standard atmospheric pressure it shows too low an altitude on high-pressure days and too high an altitude on low-pressure days. Failure to set the correct local pressure (QNH) may result in dangerous altitude errors. A 10 hPa error in altimeter setting gives 250 ft error in altitude reading.

Failure to switch on the pitot heat during take-offs in icing conditions

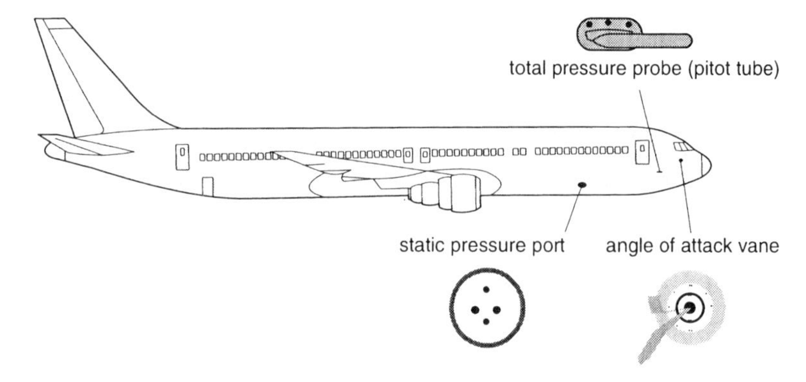

Fig. 9.7 *Sensitive air data equipment*

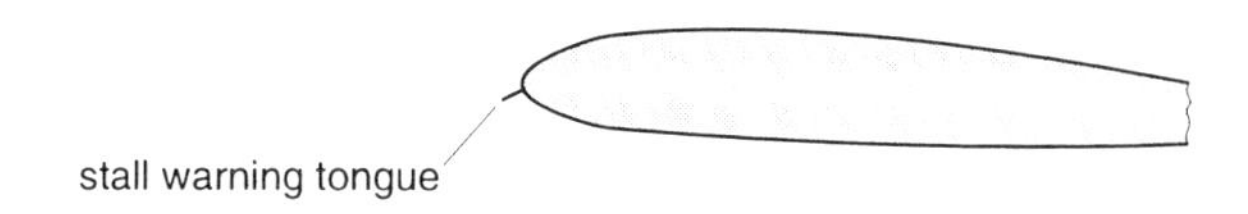

Fig. 9.8 *Stall warning device in wing nose*

has caused a number of take-off accidents. The pitot tube of a Boeing 727 froze shortly after take-off. The ice plug sealed high total pressure in the pitot-static system. When altitude increased and static pressure decreased this resulted in a false indication of increasing indicated speed. The crew increased the climb angle to prevent going too fast. As a result, the true air speed decreased and finally the aircraft stalled and pitched up into a deep stall where it remained until impact with the ground.

The crew forgot to switch on the pitot tube heat of an MD-82 taking off from La Guardia in icing conditions. At 60 knots, the air-speed indicator stopped showing an increase, jumped to 80 knots, then returned to 60 knots. The crew aborted the take-off. The aircraft overran the runway and stopped with the fuselage nose on a dyke.

Pitot tubes can be blocked even in summertime. A heavy transport aircraft took off at too high a speed and nearly blew the tyres because of a too-low speed indication. Insects had built nests in the pitot tubes during a stopover. This problem has been reported several times. In one case, the stopover time was less than one hour.

Blocked static ports have caused incidents and accidents. A heavy military transport making an I.F.R. non-precision approach nearly flew into the ground several miles from the airport. The altimeters indicated that the aircraft was too high on the approach. A check after the landing showed that the ground crew had forgotten to remove tape over the holes of the static port after washing the aircraft. It pays to use brightly coloured tape and make a walk-around inspection of the aircraft before departure.

A light twin on a non-precision approach in poor visibility and freezing rain flew into trees below the approach path a short distance from the airport. The accident investigation showed that the altimeter had indicated a safe altitude at impact. Freezing rain had probably been forced into the static ports blocking the system shortly before the crash. During the descent in icing conditions, with non-pressurised aircraft without static-port heating, it pays to make a quick check of the altimeter readings by switching to the alternative static-pressure source inside the cabin.

Damaged angle of attack vanes give false stall warnings. The Caravelle in fig. 9.9 pitched down immediately after take-off. The crew pulled up but the aircraft pitched down again. The crew made a split-second decision to crash-land on the runway instead of striking the boulders beyond the end of the runway. The accident investigation showed that faulty wiring in the angle of attack indicator system activated the stall-preventing stick-pusher at too low an angle of attack. The aircraft was flyable but the crew had the impression that it would crash.

The stall warning vanes at the wing leading edges of a light aircraft are easily blocked by ice. However, stall warners do not warn in time in icing conditions (*see* chapter 10).

Faulty air data systems pave the way for take-off and landing accidents. The systems must be protected against ice formation, insects and ground collision damage. Preflight system checks are essential for safe flight.

Fig. 9.9 *Crash caused by faulty angle of attack system*

Preflight Checks

Walk-around checks for leakages, structural damage, open hatches, damaged tyres, and inspection of fuel volumes (light aircraft) is good life

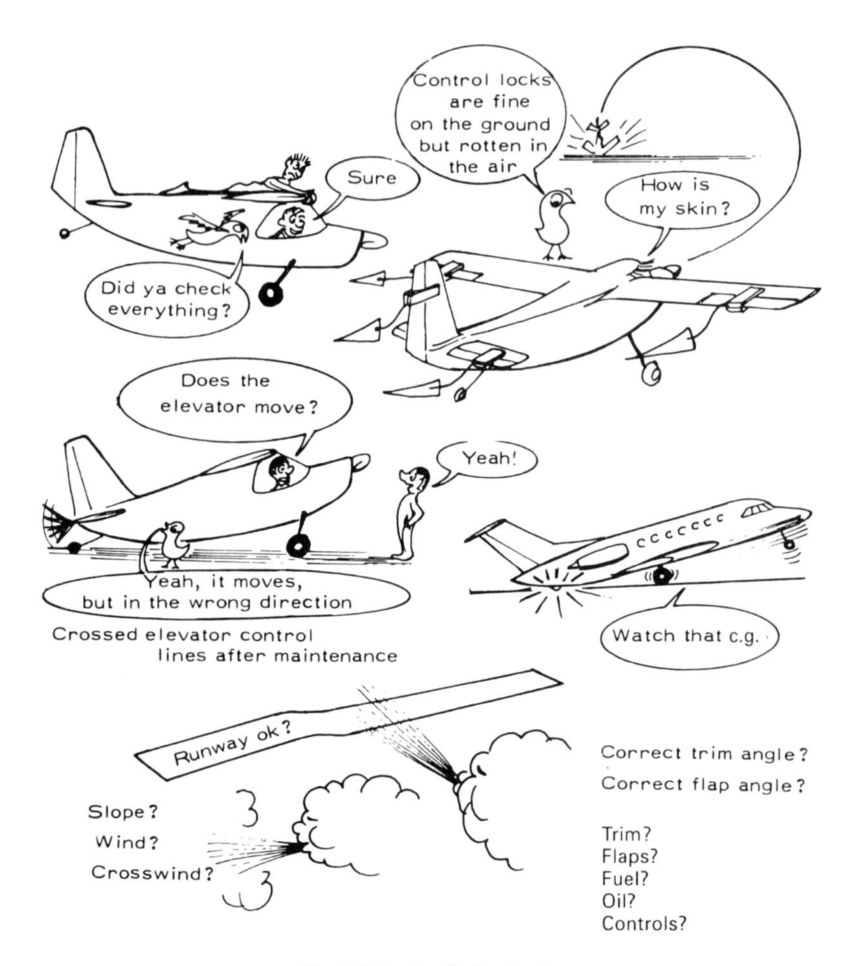

Fig. 9.10 *Preflight checks*

insurance. Furthermore, all aircraft have a preflight check list which must be carefully followed. A few misses are illustrated in fig. 9.10.

Faulty trim settings, forgetting to set slats and flaps correctly, failure to switch on air data systems, heat, etc., continually cause accidents.

Take-off

The following factors affect the take-off distance of a given aircraft:

- engine thrust, which is a function of power setting, temperature and air density (affects acceleration)

- aircraft weight (affects acceleration and lift-off speed)
- flap setting (affects lift-off speed and drag)
- runway friction and slope (affects acceleration)
- air density (affects true lift-off speed).

These factors are used in take-off distance calculations. With all data known, there is no problem making exact distance calculations. However, the take-off runway required is also affected by a number of factors.

- The aircraft postion. Is the aircraft positioned as close to the runway starting zone as possible? Every foot down the runway adds to the distance!
- Brake application during power increase; a running start may increase the take-off distance by a few per cent.
- Take-off rotation; do you rotate exactly at the assumed rotation speed and rate? One second's delay may increase the take-off run by about 20%.
- Wind, temperature, runway slope and density altitude; adverse combinations of these factors may increase the take-off roll by another 20%.

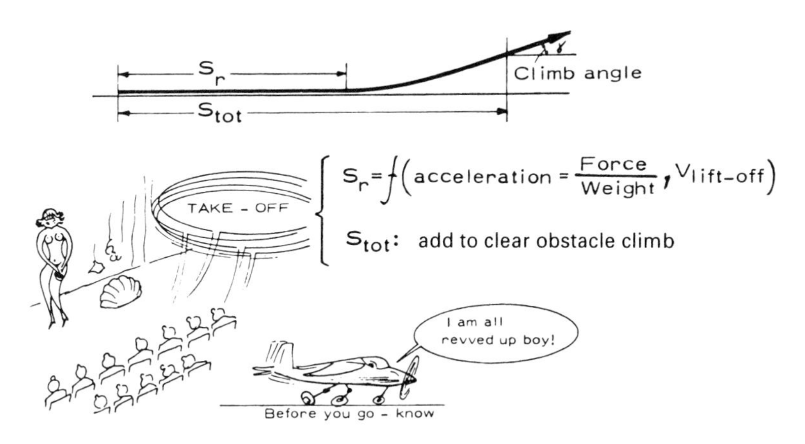

Fig. 9.11 *Take-off*

Lack of attention to details when preparing for take-off from short fields with light aircraft may create high risks especially if the take-off is attempted from a wet, soft grass field. In this case, the ground-rolling friction may be many times as high as the rolling friction on a hard, smooth surface. The total effect may be at least a doubling of the take-off distance. The pilot of the aircraft shown in fig. 9.12 landed on a short grass strip on a warm summer day, a calm day after several days of continuous rain. The landing was no problem but the take-off attempt (with a fully loaded aircraft) ended with a flip on to the back beyond the end of the runway. The rolling friction on the soft grass field was considerably higher than the friction the aircraft could cope with on the warm day. This is a fairly common summer flying accident.

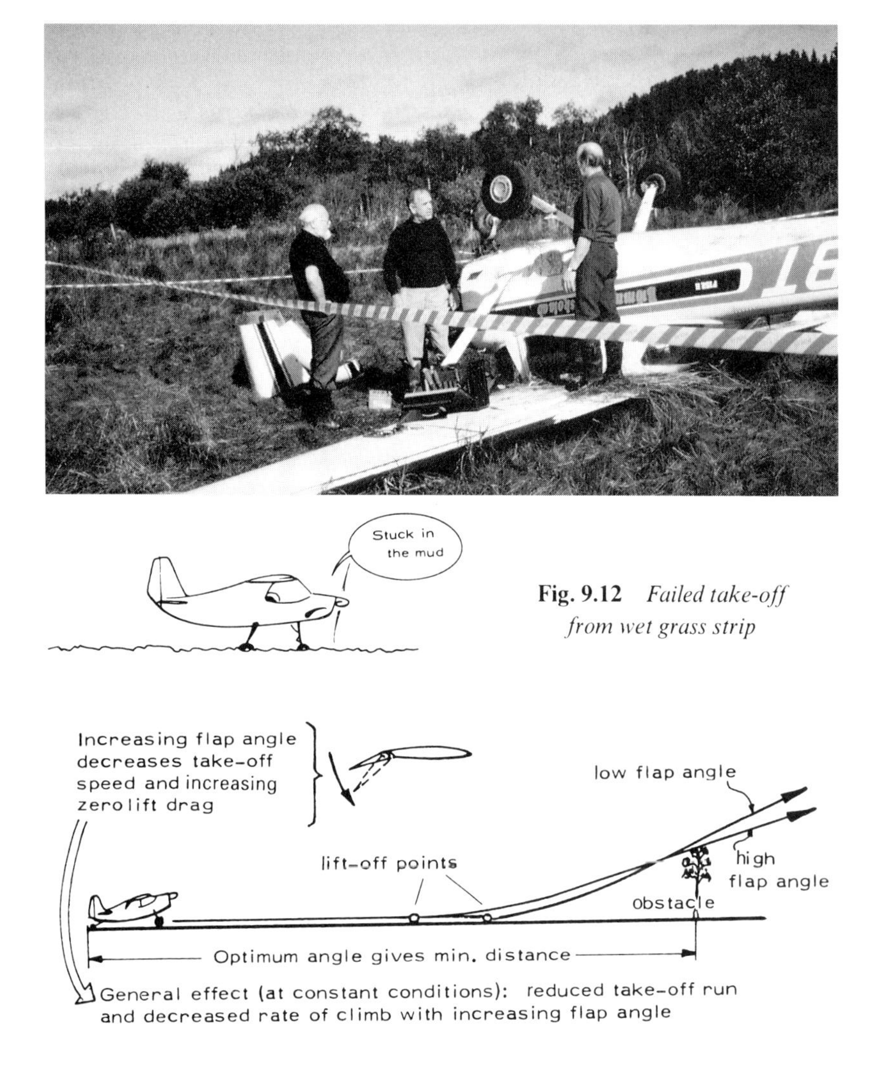

Fig. 9.12 *Failed take-off from wet grass strip*

Fig. 9.13 *Increased distance to clear obstacle with increasing flap angle*

In wintertime, the wet grass field accidents are replaced by accidents caused by friction increase due to slush or snow or drag increase due to wing contamination (*see* chapter 10).

It may be possible to reduce the take-off run on a runway with high friction by increasing the flap angle. However, acceleration and climb performance after lift-off may be so poor that obstacles in the climb-out path cannot be cleared (fig. 9.13).

It is dangerous to select flap angle based only on ground roll. Excess thrust is needed in rotation and climb-out to maintain speed and climb angle.

Too high a flap angle (landing flaps) may increase drag so much that the ground-roll-distance increases even if the lift-off speed decreases. Too low a flap angle, or failure to set take-off slats or flaps, may result in stall after lift-off. Faulty slat/flap settings is especially a problem on modern aircraft with very efficient high-lift devices.

Some aircraft have more than one take-off power setting; for example, one for normal operations and long engine-life, and one for emergency operations. For some aircraft, loss of speed in take-off may lead to rapidly deteriorating conditions (increased drag, decreased control) and, therefore, even a few per cent increase in thrust may save the day (fig. 9.14).

Failure to add available thrust has been a contributing factor in take-off accidents.

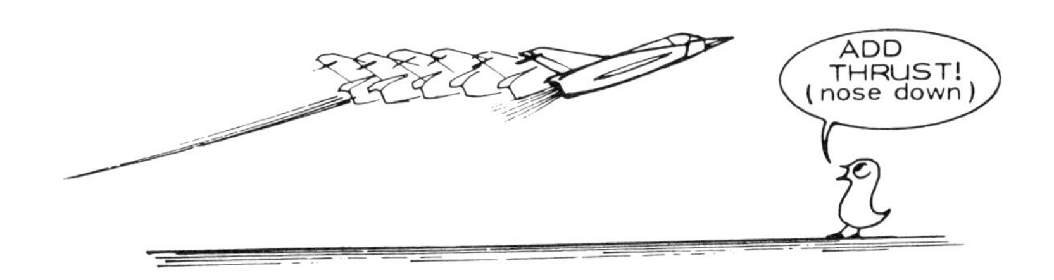

Fig. 9.14 *Add all available thrust when acceleration is poor*

Errors in take-off rotation are still happening to low aspect ratio aircraft with large lift-off angles (fig. 9.15).

High rates of change of temperature during morning hours at desert airports, and large temperature differences between thermometer locations and still air above black runways on hot days, have caused jet fighter take-off crashes and serious incidents with transport aircraft when the reported temperatures have been far below the actual ones. Fully loaded jet transports with wing-mounted fan engines scooping up hot 'runway air' have lifted off near the ends of runways and barely cleared obstacles.

For multi-engine aircraft in commercial operations it is necessary to

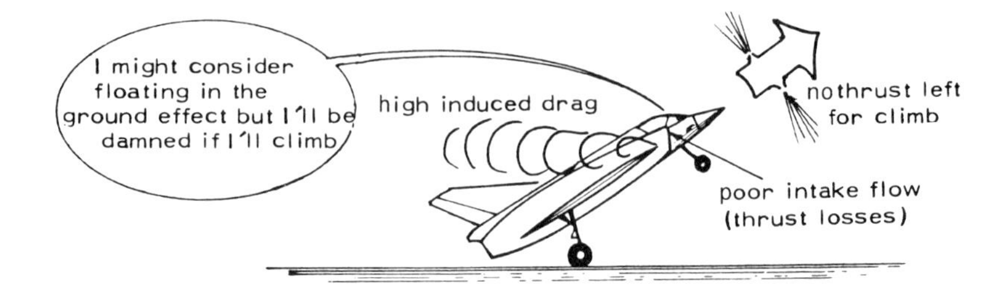

Fig. 9.15 *Overrotation at lift-off*

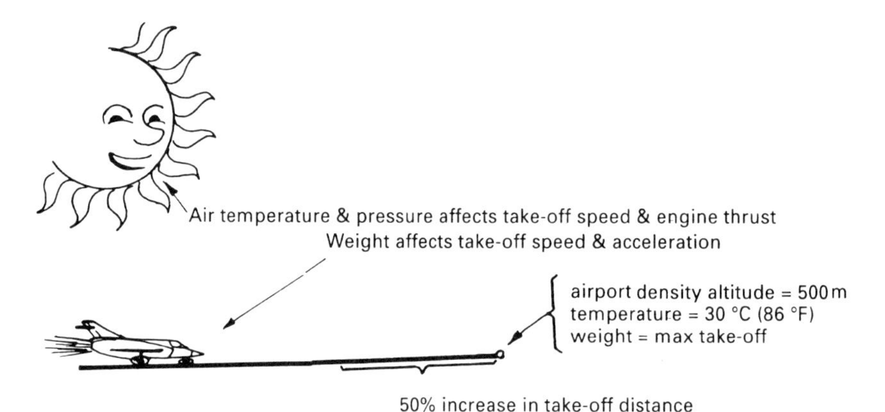

Fig. 9.16 *Danger with too low a reported temperature*

consider single engine failure in the take-off calculations. In this case, an engine failure speed (V_1) is determined. If the engine fails at this speed, the accelerate – stop distance equals the distance required to accelerate to V_1 with all engines running and climb to 35 ft with one failed engine. This is called a balanced field length (fig. 9.17).

The speed at 35 ft must not be less than 1.15 to 1.20 times the stalling speed. When the engine fails, the pilot has two seconds to react and start braking. Use of thrust reversers is not permitted since reversing may not be possible if one engine fails.

The accelerate–stop and accelerate–climb distances are flight tested by highly trained test crews flying new aircraft on good test days. The braking tests are made on dry runways with good braking friction. The test data are corrected to standard day conditions. Most transports are rotated for lift-off at a speed V_R a few knots higher than V_1 and reach V_2, a speed with a safe margin to stall with clean wings and no tailwind shear, shortly after lift-off. The balanced field length with callout of V_1, V_R and V_2 provides transport pilots with the best available cues for safe take-offs in case of

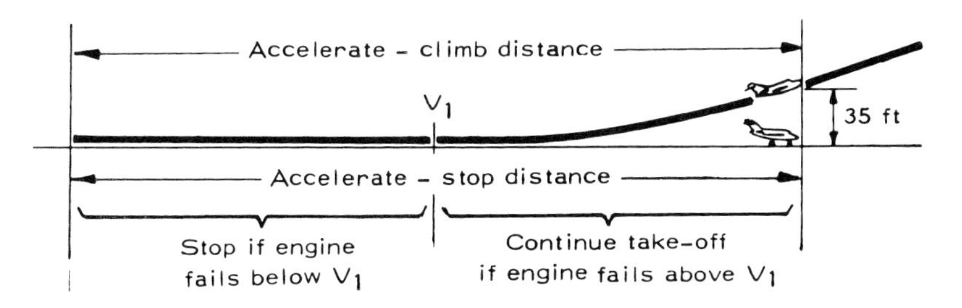

Fig. 9.17 *The balanced field length*

engine failure. However, for the following reasons it does not guarantee safety.

- The distance required to accelerate to V_1 is affected by the distance used to line up for take-off, runway up-slope, soft tyres, dragging brakes, runway contamination, aircraft surface contamination, crossed controls, engine contamination and wear, and local air temperature over the black runway surface, even if the take-off is corrected for aircraft weight, density, altitude, and tailwind.
- The distance required to climb over the 35 ft obstacle is affected by wing contamination (*see* chapter 10), engine wear, and time to identify problems and react to them and, for instance, to retract landing-gear.

Furthermore, it is much more difficult for a line crew caught by a surprising engine failure to manage a climb-out with one engine failed during severe operational conditions than it is for a highly trained test crew to make a certification test for which they are fully prepared. Still, if the take-off is runway limited it may be considerably safer to continue the take-off than to try to stop for the following reason.

The accelerate–stop distance assumes that the runway is dry and that the aircraft will stop with the nose wheel at the runway end. If the runway up to V_1 is contaminated and the acceleration distance to V_1 increases the aircraft would, therefore, run off the end of the balanced field runway even if braking was perfect beyond the V_1 point. With a contaminated runway there is no way of stopping at a balanced field length. The aircraft may run off the end of the runway with speeds up to 80% of V_1 (*see* chapter 10).

Even if take-off conditions are perfect, attempts to stop on the runway may fail for the following reasons.

- Action delayed by one second increases the stopping distance by a couple of hundred feet at the same time as the aircraft's speed increases by a few knots.

- If the runway has a hump in the middle and slopes down at the end, the braking distance increases.
- Airline crews are not used to maximum braking.
- Aircraft brakes may fail during maximum braking at maximum take-off weight. If the brakes are hot due to previous braking they may fail at any take-off weight.
- If the take-off is aborted for reasons other than engine failure, the aircraft may accelerate to speeds considerably higher than V_1 before braking is initiated.

Fig. 9.18 *A dry runway overrun*

The following overrun cases show what may happen.

- The aircraft shown in fig. 9.18 was taking off from a long dry runway in a 45° crosswind from the left. The balanced field length under the prevailing conditions was approximately 5 000 ft. When the captain tried to rotate the aircraft for lift-off nothing happened; the aircraft continued straight ahead, accelerating with all three engines running at take-off power. The crew had forgotten to switch on the hydraulic control system.

  When the crew decided to abort the take-off the aircraft had accelerated from a V_1 of 243 km/h to 265 km/h. When full braking including lift dumpers was applied the balanced field length had been exceeded. Attempts to reverse at high speed lifted the aircraft nose and it began to drift off the runway. Reversing was discontinued. The aircraft stopped in soft ground on the shallow down-slope beyond the runway end, 8 600 ft from the start of the take-off run.

- The burning transport in fig. 8.14 was stopped at the end of the runway after bird-strikes at a speed below V_1 and with a low take-off weight.
- The take-off of the burning DC-10 in fig. 9.19 was aborted at V_R due to high vibrations caused by nose gear tyre failure. A very large increase in vibration level was experienced as the nosewheel was lifted off the ground. The aircraft was at maximum take-off weight.

  The fire erupted when the aircraft's wing struck a concrete hut on the right side of the runway extension. Without the fire, the accident would not have been the catastrophe it turned out to be.

Fig. 9.19 *Fire due to collision with concrete hut in overrun zone*

The problem of designing runway ends for the safest possible overruns is a serious one. Overruns result in disasters when aircraft run into rough terrain beyond the end of the runway or collide with obstacles erected beyond the end.

The problem can be reduced if antennas and huts containing I.L.S. equipment are designed to collapse when struck by aircraft. This is the case with the antennas struck by the aircraft in fig. 9.20. The aircraft was repaired in a few days.

Take-off requires the highest mental preparation for making quick decisions to continue take-off or abort it in case of problems near V_1.

Fig. 9.20 *Collision with collapsible structure during overrun*

Climb

Aircraft climb performance is affected by:

- engine thrust (power) and thus ambient temperature and pressure
- aircraft weight
- aircraft drag (aircraft configuration)
- flight speed (since drag and thrust change with speed).

The maximum climb angle is obtained at the minimum drag speed when $T - D$ is largest, where T is the thrust and D is the drag. Since $T - D$ does not change very much near the minimum drag speed it pays to increase the climb speed in a tailwind (i.e. to reduce the climb time) and to decrease the climb speed in a headwind in order to get the best climb angles. Tailwinds reduce the climb angle and headwinds increase the angle.

The maximum climb rate (the highest vertical climb speed) is obtained at a flight speed higher than the minimum drag speed since $T - D$ decreases more slowly than the speed increases up to an optimum speed when $(T-D)\,V$ reaches a maximum, where V is the flight speed.

Even the maximum rate of climb has a fairly flat optimum which means that the rate will decrease fairly slowly as the optimum speed is exceeded. Since the climb angle then decreases, the climb distance increases. Thus, climbing at a slightly higher than optimum speed may decrease flight time and increase range.

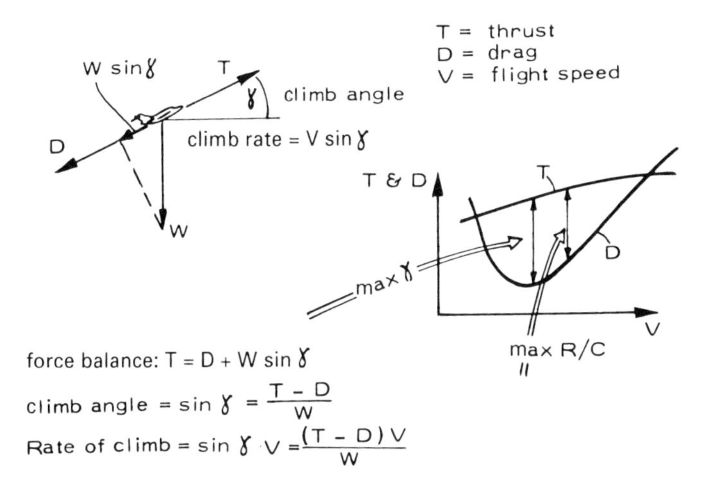

Fig. 9.21 *Climb performance*

The indicated air speed for maximum climb angle and maximum climb rate is constant up to altitudes where aircraft run into the transonic drag rise. Above this altitude, the climb Mach number becomes constant (fig. 9.22).

The true air speed increases during climb, as long as the indicated air speed is constant. However, the climb angle decreases with increasing altitude due to decreasing engine power (thrust) with decreasing air density. Finally, an altitude is reached where further climb is impractical since it becomes impossible to maintain that altitude in cruise in turbulence or to turn without losing altitude. A certain amount of excess thrust at continuously available power is required for manoeuvres.

Transports taking off must be able to climb safely with one failed engine. The loss of an engine may result in a dramatic loss of climb performance. The thrust required to overcome drag is high, especially with landing-gear and high-lift devices extended, and the excess thrust available for climb becomes small. For a typical twin-engine transport the maximum rate of climb decreases from approximately 3000 ft/min to 400 ft/min (fig. 9.23).

Some light twin-engine aircraft with reciprocating engines cannot maintain level flight with gear and flaps down if single engine failure occurs. In order to prevent stall, the nose must be lowered, the propeller feathered, the gear retracted and flaps eased in as soon as a safe speed is reached. Gear retraction immediately after lift-off and acceleration to safe flapless speed is the best insurance for optimum climb performance if single engine failure occurs after take-off.

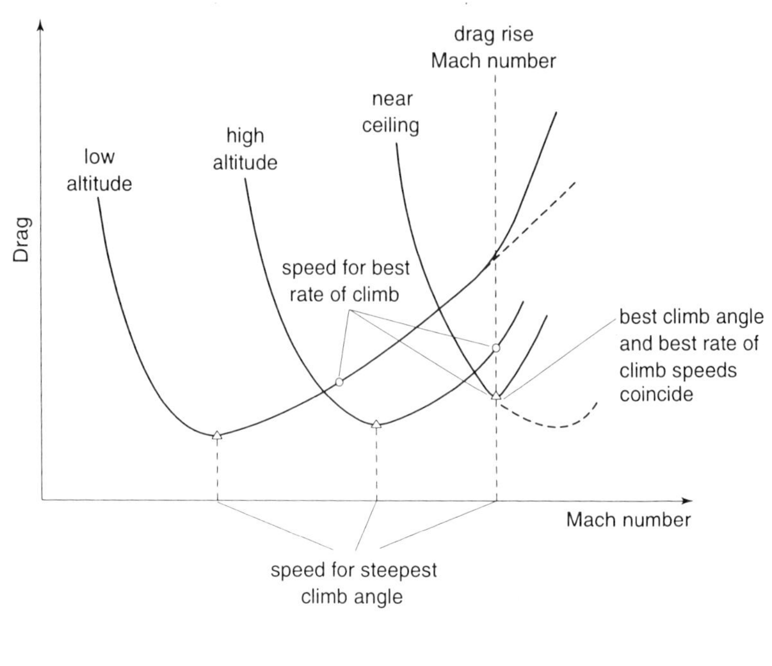

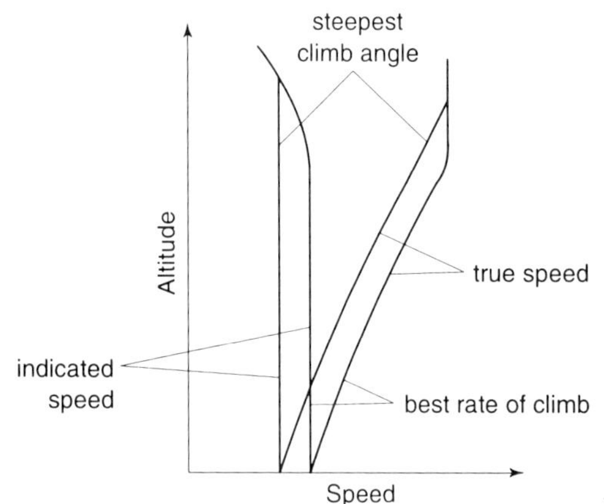

Fig. 9.22 *Effect of altitude on the optimum climb speeds*

With the low rate of climb available with one failed engine, climb performance is easily affected by wind shears, downdraughts, errors in air temperature and pressure data, and aircraft surface contamination. Frost on a wing may make climbing impossible if one engine fails (*see* Chapter 10).

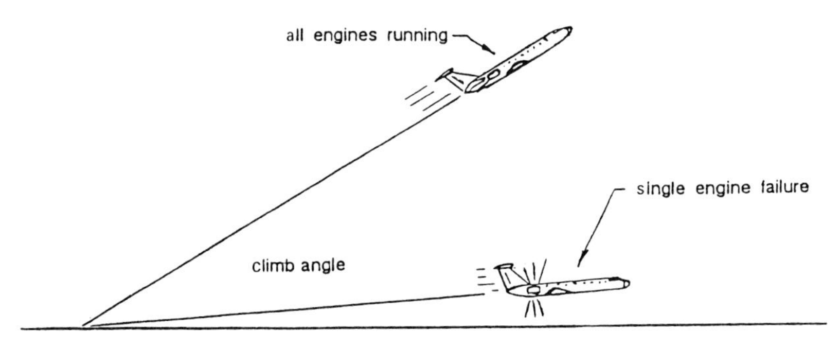

Fig. 9.23 *Reduction in climb performance with one failed engine*

A low thrust or power reserve for manoeuvring with one failed engine may make climbing turns without losing speed impossible. For some aircraft, especially old ones that are heavily loaded it may be impossible to make level turns in flight. In case of engine failure during take-off climb, it may be necessary to accelerate and climb straight ahead for some time before attempting to turn in order to have speed and altitude reserves. Steep turns after engine failure during take-off accounts for many stall-spin accidents.

Poor climb, acceleration and turning performance after single engine failure may make it impossible to climb out of an airport surrounded by mountains. A pilot may find himself 'boxed in'. Climbing into a headwind may change into climbing into a downdraught as a mountain is approached (fig. 9.24). Thus a headwind is no guarantee of a safe single engine climb over a distant obstacle (fig. 9.24). In this case it is wise to look for a suitable place to crash-land in case of engine failure. A controlled crash is always safer than an uncontrolled stall–spin.

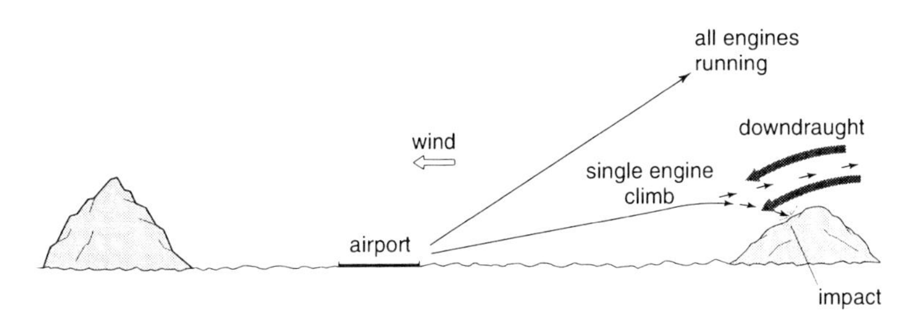

Fig. 9.24 *Headwind changing to downdraught*

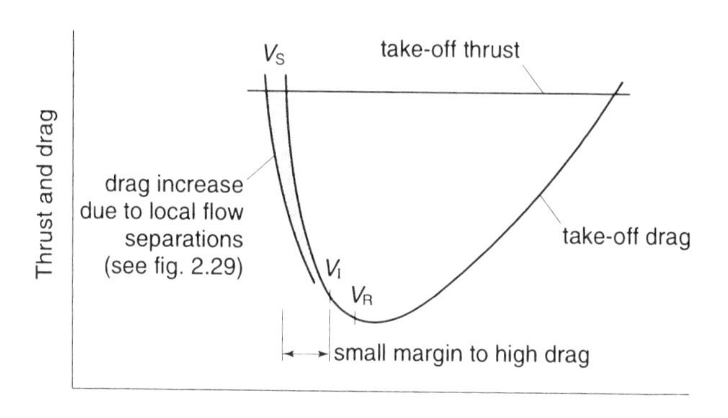

Fig. 9.25 *Typical take-off drag and thrust for a swept-wing jet transport*

Fig. 9.26 *Outclimbed by the ground*

Wind shears were discussed in chapter 7. However, observe that for swept-wing jet transport aircraft, drag may increase rapidly at speeds below V_I (fig. 9.25). Thus, in a thunderstorm with large headwind to tail-wind shears, an aircraft may end up in a semi-stalled condition with little or no excess thrust available for continued climb. Since dive to increase speed is not advisable at low altitudes the best advice that can be given is this: do not attempt take-offs below thunderstorms even if you are sure no engine will fail!

If you have taken off below thunderstorms without running into problems you have not gained good experience. You have just been lucky!

Climbing up through valleys to pass a mountain range with light, single-engine aircraft may develop into an impossible trap when the pilot discovers that the ground climbs faster than the aircraft. This happened to the pilot of the float-equipped aircraft in fig. 9.26.

Barging ahead blindly during take-off and climb, without careful pre-flight preparation, is a sure way to disaster.

Flight Envelope

Typical flight envelopes for subsonic and supersonic aircraft are shown in figs. 9.27 and 9.28.

For subsonic aircraft the minimum speed is usually determined by stall (or poor handling qualities near stall). The maximum speed is limited by thrust and drag except for high-speed transport aircraft where a dynamic pressure limit may be introduced. Jet transports never operate at maximum speed at low altitudes. It would, therefore, be a waste to design the structure for the high dynamic pressures at low-level high-speed flights.

Light aircraft are designed for maximum dynamic pressures obtained in dives. Exceeding the design dive speeds may overload the aircraft structure and increase the risk of aeroelastic bending of wing and tailplane structures.

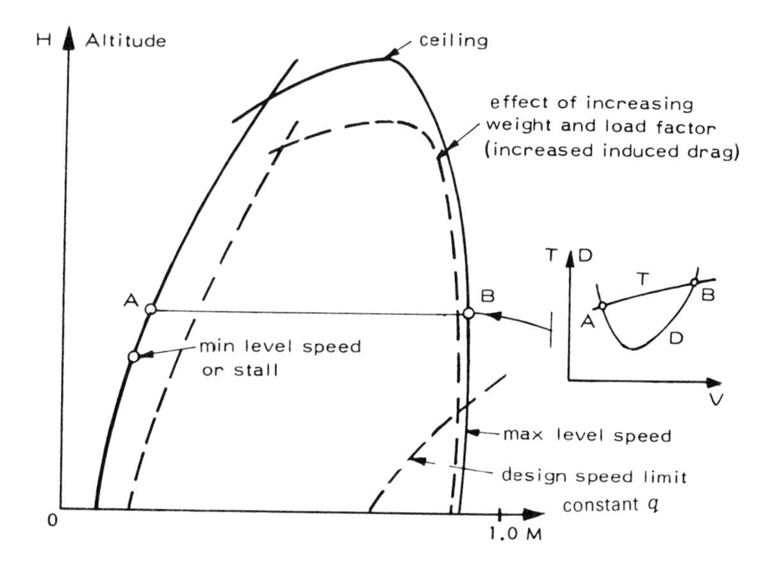

Fig. 9.27 *Subsonic jet transport flight envelope*

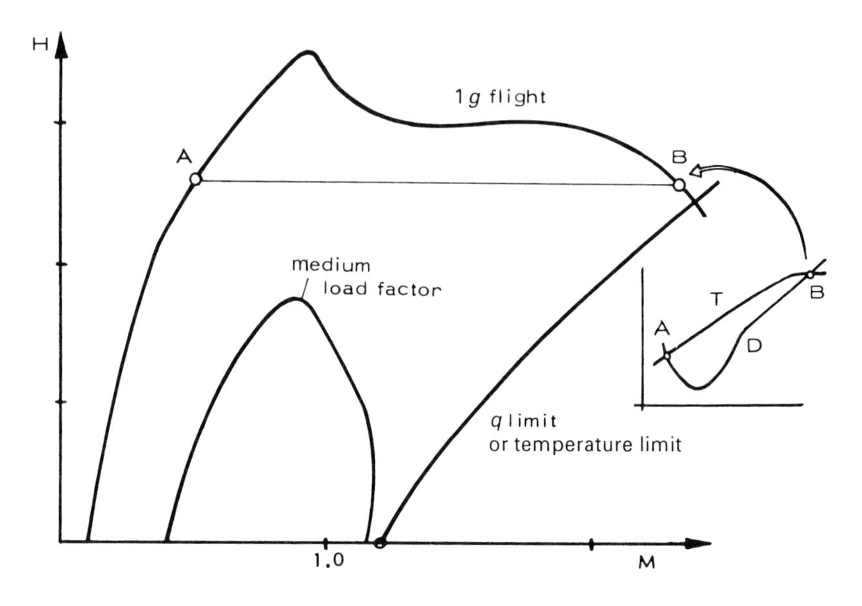

Fig. 9.28 *Supersonic aircraft flight envelope*

The risk of flutter also increases, especially if the aircraft has a sloppy control system or if the elevator mass balance has changed due to ice on or inside the elevator. Stay away, with good margins, from the maximum speed limits.

A jet transport can exceed the low-level structural design speed limit in level flight but with a functioning air data system, standard take-off and climb to cruise altitude and speed warnings, the risk of any crew exceeding the structural limit in level flight or climb is small.

At high altitude, jet transports may run into Mach limits due to the transonic flow problems described in chapter 3. A fully loaded DC-8 on a high-speed descent on autopilot ran into a sudden headwind shear. The increased Mach number resulted in a shock-stall at the outboard wing sections. At the same time, the aircraft's tuck-under compensator (which gives a nose-up trim to compensate for transonic tuck-under) reacted to the increased Mach number and gave an automatic nose-up trim. This, together with outboard wing stall, resulted in the autopilot disconnecting and a 2 *g* pull-up. The pilots threw themselves at the controls, overreacted and sent the aircraft into a –0.5 *g* bunt. Dinner had just been served. It hit the ceiling in the passenger cabin.

In spite of the risks of flying at the Mach limit, I have flown with crews using the Mach warning 'rattle' as a speed indicator for high-speed let-downs. Incompetent lack of risk awareness!

The minimum speed of low aspect ratio supersonic aircraft is usually limited by high induced drag and the need for excess power for acceleration in case of aborted landings. However, poor low-speed handling qualities may also limit the minimum speed.

At supersonic speeds, maximum design dynamic pressure or maximum allowable temperature may determine the maximum speed. At high altitudes the thrust and drag curves may be nearly parallel giving high equilibrium speeds but low acceleration power to reach those speeds. In a case like this, the aircraft will decelerate to subsonic speed at fairly low load factors.

Endurance and Range

Fuel consumption and speed are two primary factors in endurance and range determinations (fig. 9.29).

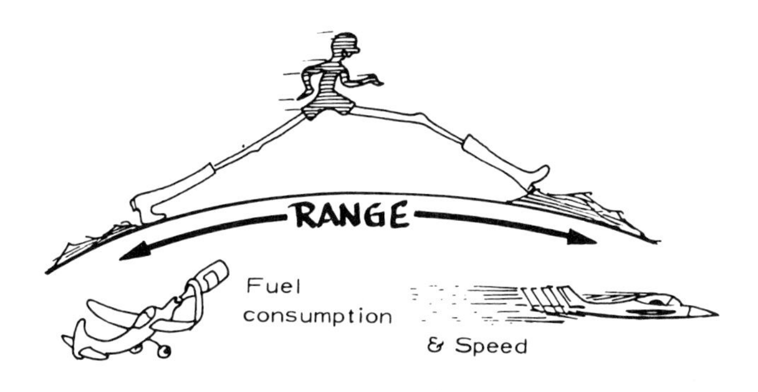

Fig. 9.29 *Range*

The reliability of endurance and range data is determined by the reliability of the aircraft's drag data, the installed engine thrust (power) and the fuel consumption data. The calculated performance is checked in flight tests. The test aircraft are new and any operational wear affecting aircraft drag and engine performance reduces endurance and range.

The maximum endurance is obtained at the minimum-drag speed, i.e. at the indicated speed for maximum climb angle (fig. 9.30a).

Since the indicated speed is constant with increasing altitude the true air speed increases.

The maximum-range speed is higher than the maximum-endurance speed since the fuel consumption per mile decreases as speed increases (even if the fuel consumption per hour increases) up to the speed where fuel consumption per hour divided by true air speed reaches a minimum.

For general aviation aircraft flying at altitudes below 5000 ft the effects

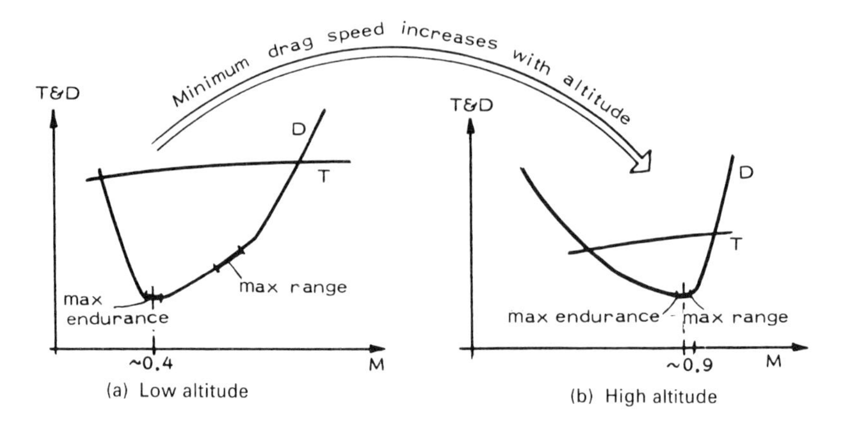

Fig. 9.30 *Endurance and range*

of altitude on range and endurance are small. For jet transports, on the other hand, the true air speed increases by around 45% between sea level and normal cruising altitudes, while the engines' fuel consumption, at constant thrust, decreases. Jet engines are designed for optimum cruise performance at the high r.p.m.s required for high-altitude flights. The high-altitude range performance of jet transports is nearly twice as good as the range at low altitudes.

At altitudes above that where aircraft reach transonic drag-increasing Mach numbers, the indicated speeds for maximum endurance and range decrease and approach each other (fig. 9.30b). Since minimum drag increases rapidly above this altitude, both endurance and range decrease.

Both endurance and range change slowly near the optimum speed and altitude. For this reason, it may not be practical to fly exactly at the optimum speed. At the minimum-drag speed any loss of speed may result in further loss of speed or altitude when drag becomes higher than thrust. It is safer to fly at a higher speed where a loss of speed gives acceleration and an increase in speed a deceleration (fig. 9.31).

Range is not usually a safety problem for jet transports carrying sufficient fuel reserves. However, attention to altitude winds pays off. Running into a 100-knot jet stream decreases range by about 20% while flying in a 100-knot tailwind at a lower altitude increases range by 20%. Thus, wind awareness is important. General aviation pilots are not always as alert as they should be. Negligent preflight fuel checks, flights on too rich a fuel–air mixture and insufficient *en route* fuel checks have resulted in emergency landings on unprepared fields with nose-overs (fig. 9.32).

Wind effects on slow general aviation aircraft are larger than on high-speed jet transports. An aircraft cruising at 100 knots loses 20% of its range

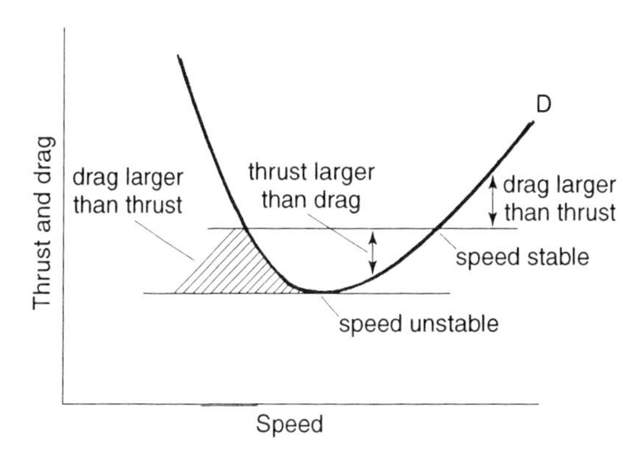

Fig. 9.31 *Speed stability*

in a 20-knot headwind. Since meteorology is not an exact science and the effects of wind changes may be large, it is obviously good life insurance to check the amount of fuel actually being used against planned consumption at reasonably frequent intervals *en route*, to update wind information, and to land and refuel if consumption begins to eat your reserves.

Fig. 9.32 *Dry-tank emergency landing with nose-over*

In several of the light twin accidents I have investigated, a major contributing factor has been high stress due to the aircraft being extremely low on fuel combined with unexpected problems during landing approaches. Too many people have been killed by taking unnecessary risks. The pilots had plenty of opportunity to refuel *en route*.

The following factors may have considerable effect on fuel consumption:

- leakage from pressurised cabins through damaged door seals
- poorly rigged control surfaces and flaps (too high or too low)
- floating spoilers or air brakes
- leakage from hot-air ducts into the wing's leading edge airflow
- poorly maintained pneumatic de-icing boots
- wing and empennage leading edge damage (erosion and dents)
- aircraft surface contamination (especially at wings) such as frost, snow and ice in wintertime, and dead insects in summertime
- heavy *en route* turbulence
- prolonged flight through heavy rain.

The following is a typical general aviation dry-tank accident. The pilot had to land in the ocean due to lack of fuel after a two-hour flight over water. The aircraft was reported missing and the search-and-rescue helicopter found the pilot sitting on a rock. 'I had plenty of fuel for the return flight when I landed last night,' he said. 'Someone must have stolen fuel during the night'. Checking the fuel would have saved an expensive aircraft. It still rests on the bottom of the Baltic Sea.

Practical Problems of Stall

Stall was discussed in chapters 2 and 4, and the reduction of C_{Lmax} with increasing Mach number was shown in chapter 3. The basic stall requirement is that aircraft should be fully controllable in roll, yaw and pitch down to stall, should have good stall warning characteristics (buffeting) warning devices (horn or stick-pusher) and should pitch down gently at stall.

This requirement is easy to meet with fairly straight wings with twist or wing-root stall strips. But as wing sweep increases it becomes impossible to avoid wing-tip stall and pitch-up. Delta wings always pitch up at high angles of attack.

Swept-wing transports with fuselage-mounted engines and long fuselage sections in front of the wing may pitch up to high angles of attack and be locked into deep stall by the downwash on the T-tail tailplane by the fuselage vortices (fig. 2.25).

However, deep stall can be delayed or prevented by increasing the span of the tailplane. This way the tailplane tips extend into the upwash part of the body vortices. The lift thus obtained on the outboard sections of the tailplane counteracts the down load on the inboard area created by the vortex downwash (fig. 9.33).

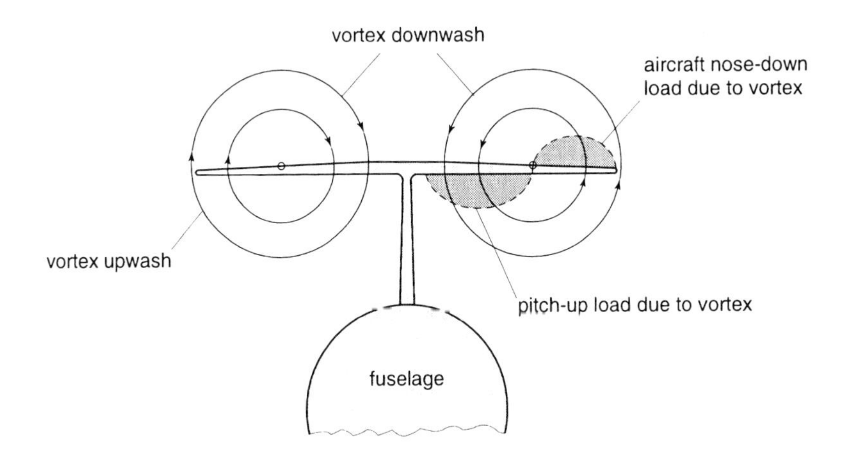

Fig. 9.33 *Effect of increasing tailplane span on deep stall*

It is also possible to delay deep stall by mounting strakes on the sides of the fuselage nose (fig. 9.34). The strakes generate vortices trailing low along the fuselage sides. They pass below the T-tail stabilizer up to high angles of attack and delay the formation of vortices on the fuselage top. Thus, deep stall is delayed. However, stay away from deep stall. There is no good flight-tested method for recovery!

During stall tests, engines are throttled back to zero power or thrust in order to eliminate the effects of slipstreams and the vertical thrust component on lift. For propeller aircraft with approach power, this means that the stalling speed during landing is lower than the test value due to the lift created by the deflected slipstream.

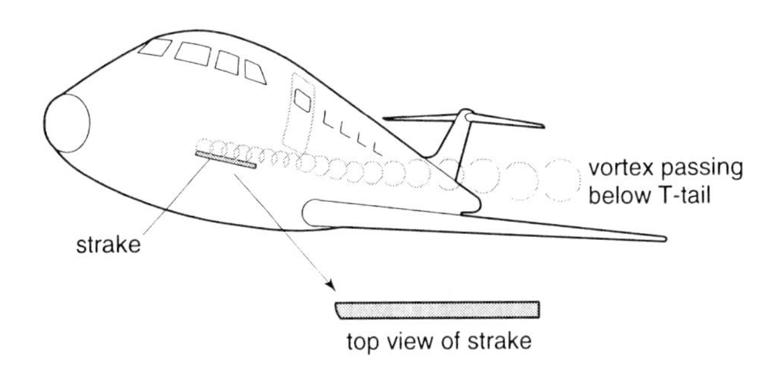

Fig. 9.34 *Fuselage strakes*

Special problems are encountered in stall testing of jet transports. Due to high drag as stall is approached, the aircraft decelerate quickly in horizontal flight with zero thrust. Since it takes some time for a wing to stall the high pitch rate required to maintain level flight, as drag goes up and deceleration increases, a dynamic maximum lift coefficient is obtained. This may be considerably higher than the steady state value. For this reason the deceleration when approaching stall in flight tests is limited. In order to maintain slow deceleration the aircraft is flown in a bunt as stall is approached. The bunt reduces the load factor and the flight test stalling speed is therefore lower than the speed obtained in steady-flight during a low-powered approach. Corrections for the reduced load factor are not necessarily made. This, together with the lack of slipstream effect on lift, gives the jet transport a lower margin to the published stall than that of propeller-powered aircraft with wing-mounted engines. Also, due to the drag increase of swept-wing transports as stall is approached, the drag at stall may be higher than the maximum power available (fig. 9.25).

The minimum speed of aircraft with highly swept wings and aircraft with delta wings is usually not limited by stall. Instead, high drag, and possibly stability and control problems, limit the minimum speed. A typical requirement for a fighter is that it must be able to accelerate at least 1 m/s² without afterburner in level flight at maximum landing weight. Without an acceleration requirement, there is a considerable risk of speed loss to a high-drag condition from which recovery is not possible at low altitude since drag exceeds maximum available thrust.

Stall testing of transport aircraft is difficult. Stall is affected by turbu-

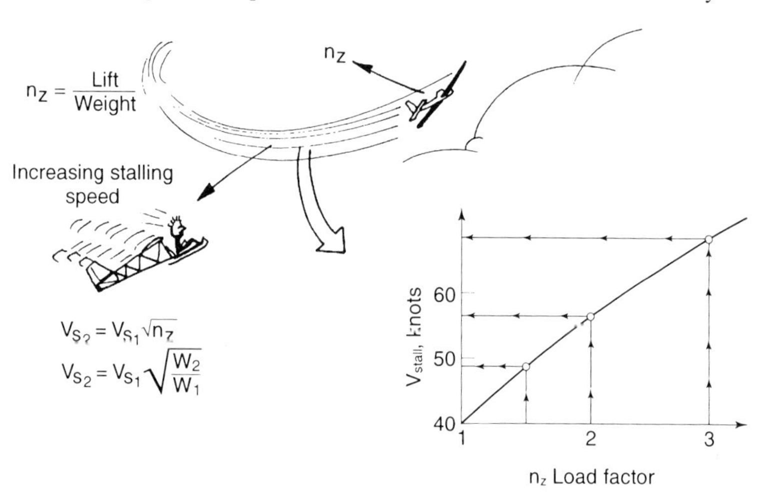

Fig. 9.35 *Effect of load factor on stalling speed*

lence, control surface movement, wing leading edge erosion, dents, dead insects, etc. Specially trained crews and test equipment are required. For this reason post-maintenance flight tests should never be made at speeds below which stall warning occurs.

Since the stall angle of attack for a given flap angle is roughly constant at low speeds, the stalling speed will increase with the square root of the load factor or the weight ratio (fig. 9.35). Thus, a load factor of 2 increases the stalling speed by approximately 40%, and a 10% increase in weight increases the stalling speed by 5%. Hence, an aircraft which stalls at 40 knots in level flight stalls at 47 knots in a 1.5 *g* turn and at 56 knots in a 2 *g* turn.

If a light aircraft which stalls at 40 knots, runs into a 3 ft/sec vertical gust with a wing-tip, the tip may stall at about 45 knots (fig. 9.36a). If an aileron is moved downward to stop the tip from moving downward when it stalls, the stall will be exacerbated since the downward movement of the aileron lowers the stall angle even if it increases lift at lower angles of attack. This means that there is no guarantee that an aircraft designed for wing-root stall in still air will not stall–roll at speeds above the nominal stalling speed in turbulence.

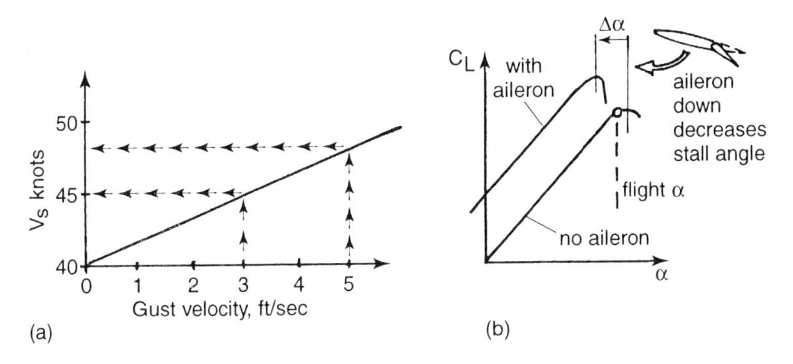

Fig. 9.36 *Effects of vertical gusts and aileron angle on stall*

Since the stalled wing has a much higher drag than the climbing wing in a stall–roll, the aircraft may enter a spin as it stalls unless the pitch control is moved forward to reduce the angle of attack as soon as the wing drops (*see* 'spin' in chapter 4).

The most common types of stall–spin accidents occur during attempts to return to the airport after engine failure, and, in the single-engine general aviation case, during low-altitude circling over points of interest (fig. 9.37).

Another fairly common type of general aviation low-altitude stall accident occurs during tight turns from fairly strong tailwinds on the downwind leg to strong headwinds in the final approach. Pulling too high a load

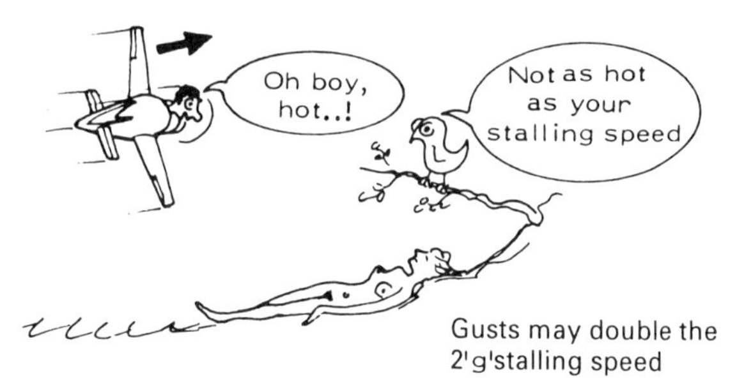

Fig. 9.37 *Low-altitude high-risk turn*

factor while trying to make a 'perfect' turn relative to the ground has brought down many aircraft.

Remember that flight manual data on stall are based on tests made with new, clean aircraft. In real operations, even small insects on a high-performance wing leading edge may stall an aircraft before the stall warning device is activated. Published data are not valid for contaminated wings (*see* chapter 10).

Descents and Approaches

During descent an aircraft reduces both altitude and speed. The mass forces due to deceleration increase the gliding distances down to the minimum-drag speed where the maximum gliding distance is obtained (fig. 9.39).

Engine failure in the approach may tempt the pilot to stretch the glide to the airport. This is not possible. It results in deceleration and stall. A controlled crash at a minimum-risk area outside the airport is much less dangerous than an uncontrolled stall elsewhere. In a case like this, the following is worth observing. A constant-speed descent with a clean aircraft gives an indication of maximum gliding distance. If the path does not point at the airport, you will not get there by stretching the glide at a reduced angle.

Small speed changes on either side of the optimum has little effect on the gliding distance due to the small drag changes near the minimum-drag speed. For this reason it pays to increase the speed a few knots in a head-wind, since decreased gliding time decreases the wind effects and decreases the speed by a few knots in a tailwind.

Transport aircraft should be established on the glide slope with approach flaps and gear down in ample time before landing. Destabilised

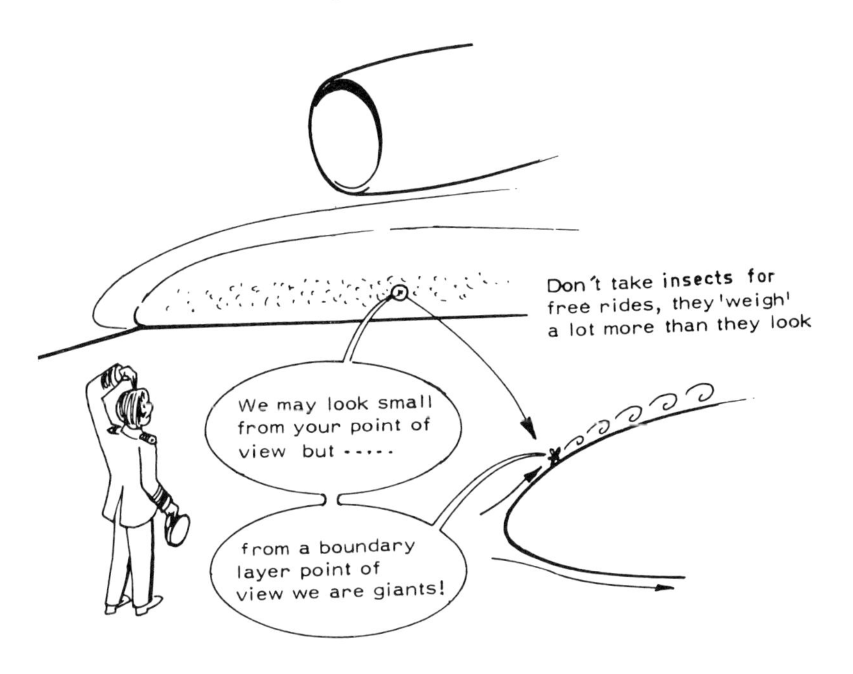

Fig. 9.38 *Insect-induced stall*

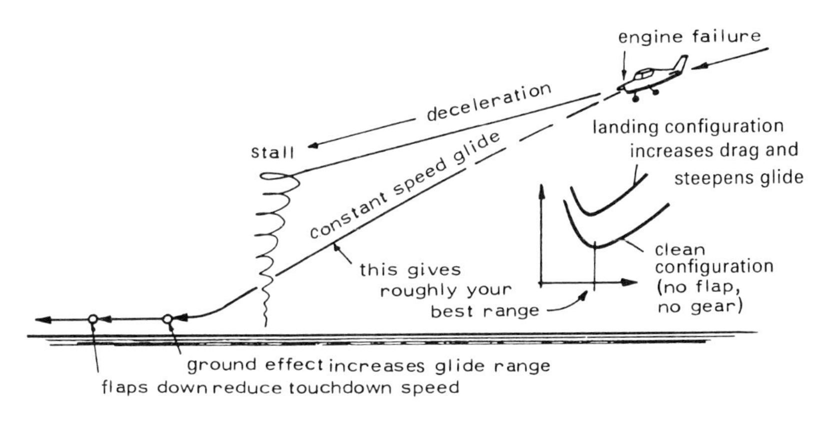

Fig. 9.39 *Glide with a failed engine*

approaches, above the glide slope at too high a speed or below the slope at too low a speed, cause crashes.

However, early extension of flaps and gear combined with premature reduction of the approach speed may more than double the landing fuel consumption compared to a higher speed, straight-in low-thrust approach. But idle-thrust deceleration requires careful speed monitoring.

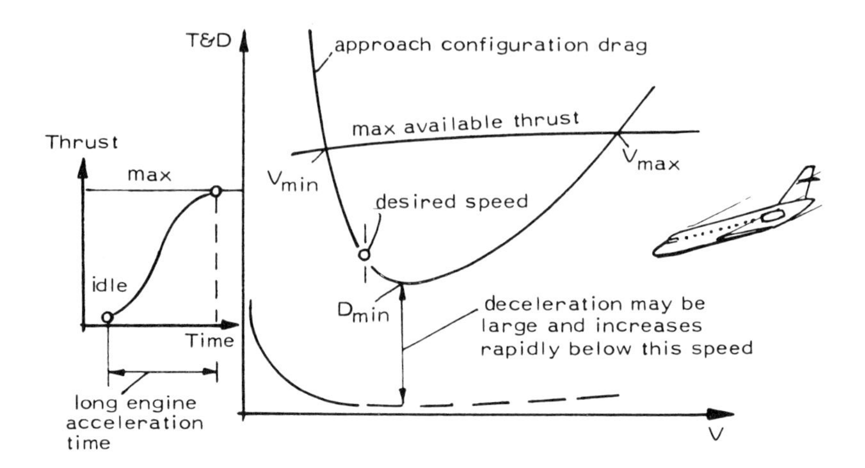

Fig. 9.40 *Deceleration in approach*

If thrust is not added in time to stop the slow-down at the desired final-approach speed, the aircraft may decelerate beyond V_{appr} (final approach speed) to critical high-drag speeds in a few seconds (fig. 9.40). The autothrottle adding thrust as the desired speed is reached solves this problem, providing it works. Autothrottle does not eliminate the need for speed monitoring.

The problem of speed and glide-path control is accentuated for low aspect ratio, highly swept fighter aircraft at low final approach speeds below the minimum drag point (D_{min}) in fig. 9.41. Attempts to use only the elevator to regain altitude when the aircraft sinks below the glide

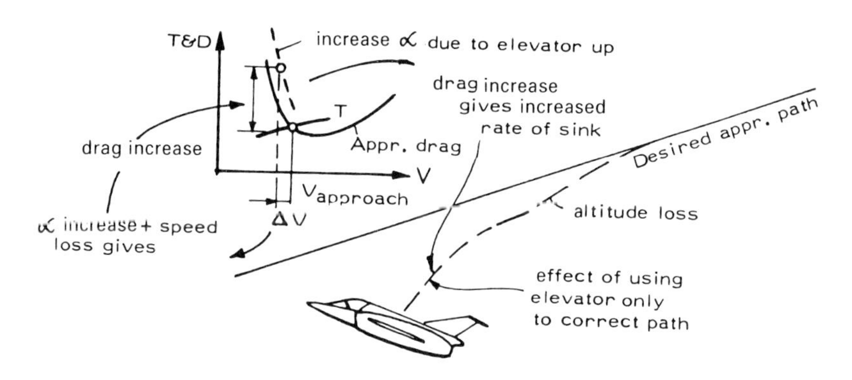

Fig. 9.41 *Effect of using only elevator for glide-slope control at speeds below* $V_{D_{min}}$

slope results in increased drag and increased sink rate. Thrust must be used for glide-slope control. Without autothrottle it is necessary to maintain considerable margins to the speeds where drag equals maximum available thrust.

The risk of ending up in high-drag attitudes if speed is lost during approach and landing explains why it is dangerous to fly into thunderstorm downbursts and wind shears, and to abort a landing and climb from a headwind into a tailwind shear (*see* chapter 7).

Three important points regarding landing in low-level wind shears are repeated below.

- An approach in a headwind along a geometrically fixed slope requires more thrust than an approach in a tailwind because the actual slope flown in a headwind is more shallow than the geometrical slope (fig. 9.42a) and in a tailwind it is steeper (fig. 9.42b).

- Depending on the shear rates, flights from tailwind to headwind shears may make the aircraft balloon and land too far down on the runway if the glide path is not monitored, or decelerate to high drag and sink through the slope if thrust is not added and carefully monitored.

  A headwind to tailwind shear may stall the aircraft or set up a high-drag sink rate through the glide slope if thrust is not added and

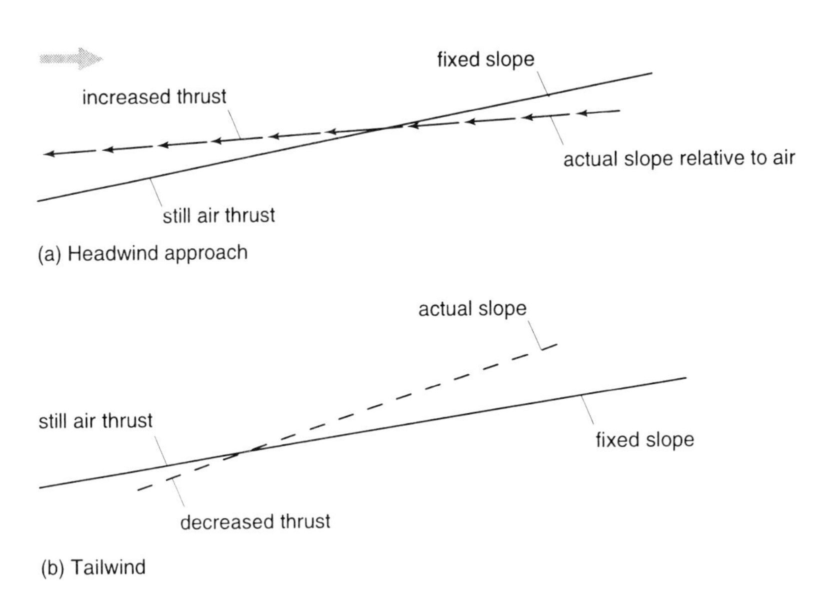

Fig. 9.42 *Thrust required in headwinds and tailwinds*

insufficient thrust is available. However, too much thrust may increase speed in the tailwind too much and result in a high-speed landing far down on the runway with the risk of overrun.

- Never make a quick climb into a tailwind from a headwind if you decide to discontinue the approach. The resulting loss of speed may be sufficient to stall the aircraft.

In one case on record, a DC-9 was on approach to a coastal airport in a reported headwind. However, unknown to the crew, the headwind was a result of very cold inland air spilling over a mountain and flowing down a valley over the airport. Above this, a fairly strong, humid tailwind was blowing from the ocean. The shear was formidable.

When the DC-9 flew into the headwind it bounced back into the tailwind, lost speed and mushed back into the headwind. It took three bounces to regain enough speed for climb-out in the tailwind. There was not enough runway left for landing.

The exact effects of wind shears are difficult to predict since they depend on a number of factors including:

- shear magnitude and rate
- aircraft speed (margin to high drag and stall)
- aircraft weight and c.g. (pitch response)
- deceleration rate along the glide slope (thrust setting, time to change thrust)
- crew response (time to change thrust, magnitude of thrust change, flight path control, etc.).

For this reason, it is obvious that the crews of transport aircraft need the best possible wind shear information and must be prepared to act as soon as the aircraft descends into the shear and go round again if necessary.

Friction between the air and the ground creates a thick boundary layer with decreasing headwind during final approach and landing flare (fig. 9.43). For low aspect ratio aircraft making approaches at low speeds this may result in high drag and high sink rates into the ground unless thrust is added as the wind decays. Attempts to pull up with the elevator increases drag and sink rate.

High aspect ratio aircraft with little or no sweep make landing approaches at speeds above the minimum drag point (D_{min}) on the drag curve (fig. 9.44). If the wind decays rapidly close to the ground, the aircraft may overshoot instead of undershooting the desired touchdown point.

To start with, the air speed may decrease even in this case. However, the decrease is counteracted by a reduction in drag as speed decreases and by

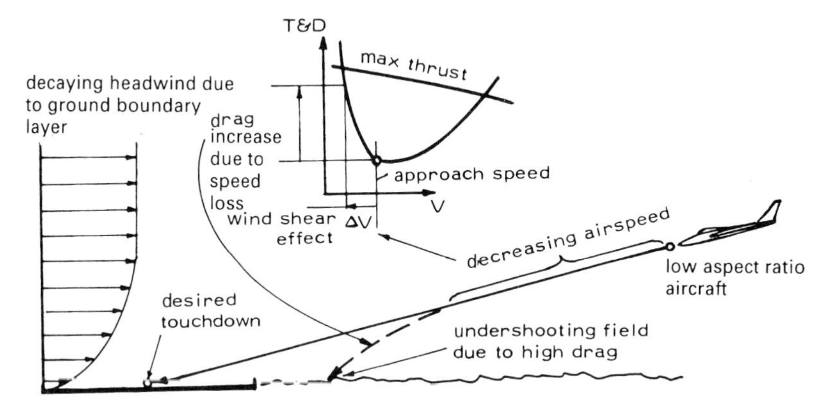

Fig. 9.43 *Effect of decaying headwind on sink rate of low aspect ratio aircraft*

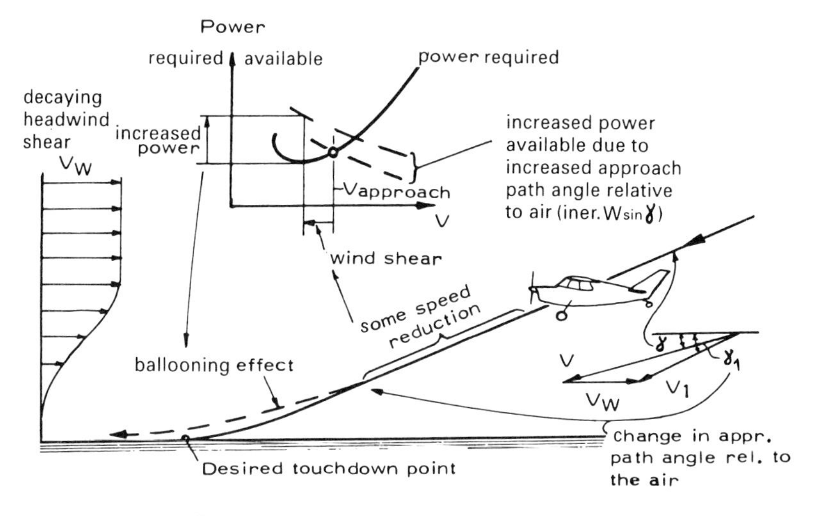

Fig. 9.44 *Ballooning in a decaying headwind*

an increase in the weight components in the flight direction as the approach-path angle (γ_1) relative to the air increases with decreasing headwind. As the low altitude with the reduced headwind is reached, the aircraft will have excess power which may result in ballooning or acceleration to a higher speed resulting in increased flare distance.

Good glide-path control requires good V.F.R./I.F.R. glide-path indication, good low-speed stability and control, and, for high performance transports, autoland systems. For supersonic fighters, autothrottle, head-up displays and a good view of the runway over the aircraft's nose are important (fig. 9.45).

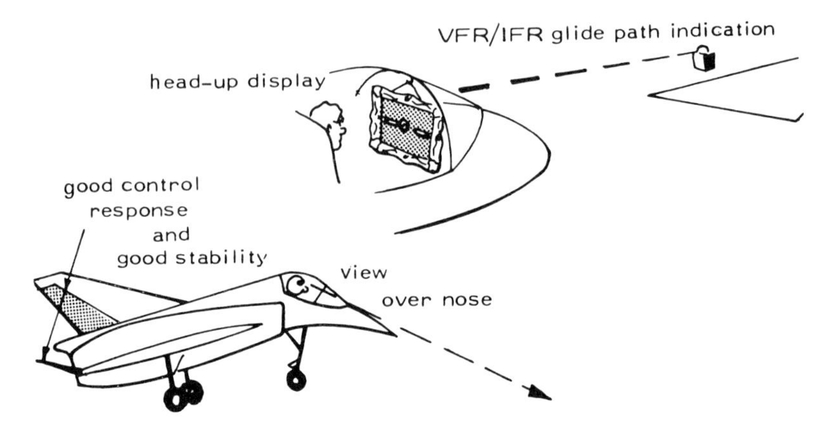

Fig. 9.45 *A few reasonable requests for good glide-path control*

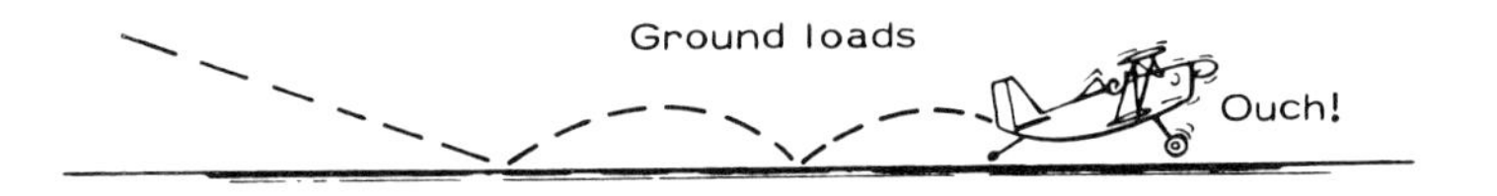

Fig. 9.46 *Hard-landing bounces*

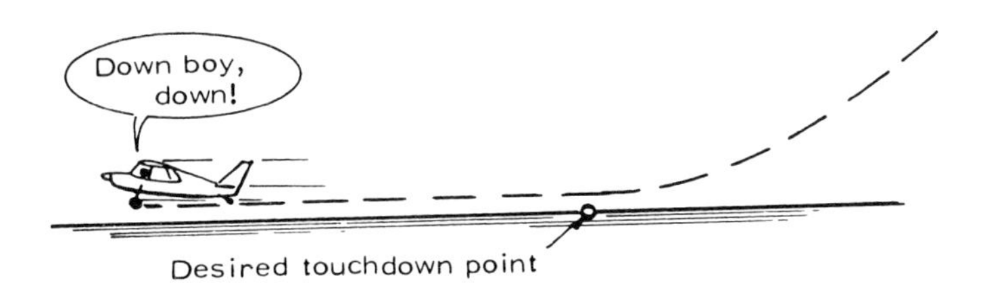

Fig. 9.47 *Hot flare in ground effect*

Nearly 50% of all aircraft accidents occur during approach and landing. During this flight phase, weather problems, reduced visibility, poor instrumentation (*see* Chapter 11), poor handling qualities, and autoland system failures may add up to give the ingredients of a 'pilot error' accident. Trouble-shooting during approach causes accidents unless at least one pilot in a transport aircraft pays complete attention to the flight progress.

Flare control contains a number of risks including:

- undershoots with risk of structural damages
- overshoots with risk of overruns

- hard landings with risk of structural damage and loss of control in bounce (fig. 9.46)
- high-speed flares over black, hot runway surfaces which can make a light aircraft float 'forever' in ground effect before touching down (fig. 9.47).

Landing a light, single-engine aircraft over an obstacle on a selected touchdown point is no problem using sideslip to control the approach path. Regarding the stopping distance, the following should be kept in mind:

- slippery runway surfaces (ice, water, wet grass) may more than double the stopping distance
- 2000 ft density altitude increases the stopping distance by about 5%
- combined effects; landing on a wet grass strip at 2000 ft density altitude in a 5-knot tailwind and a down-slope may triple the normal braking distance.

The certificated landing distances of jet transports are made over a 50 ft obstacle where the speed is 1.3 V_s to a touchdown which is sometimes called a 'controlled crash' followed by maximum braking on a dry runway surface (fig. 9.48). The distance thus obtained is multiplied by 1.67 in order to take care of various factors that may affect the landing distance. Use of reversers is not permitted in the tests.

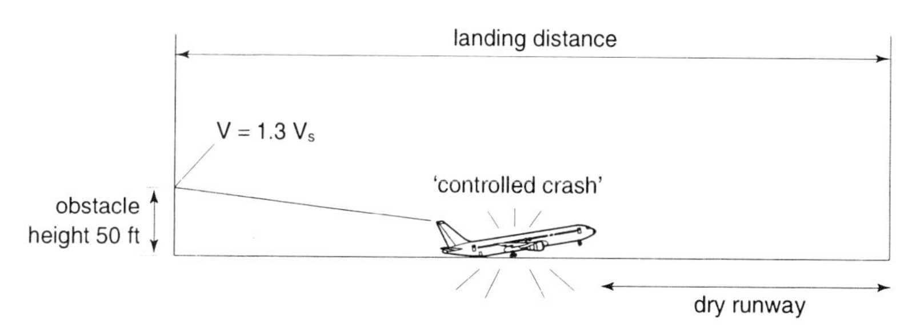

Fig. 9.48 *Landing distance flight test*

These distances have little to do with practical operational runway requirements for the following reasons.

- 'Controlled crashes' would scare passengers and quickly wear out landing gear and aircraft structures.
- Nobody can repeatedly make a final approach and flare precisely to a given touchdown point. Touchdown points for jet transports scatter by ±1000 ft. The first 2000 ft of the runway must, therefore,

be regarded as a reserve for flare and touchdown performance tolerances.

- Every second of delay in thrust reduction and brake application after touchdown may add 500–600 ft to the runway required.
- Maximum braking at every landing would be very unpleasant for the passengers and quickly wear out brakes and tyres with increasing risk of landing-gear failures and catastrophic accidents.
- Tailwind landings cannot always be prevented. A 10-knot tailwind may add approximately 1000 ft to the landing distance depending on runway conditions.
- Increased approach speed in turbulence adds to the landing distance.
- Slippery runway conditions, ice, snow, slush and water, may have very large effects on the stopping distances.

Hydroplaning makes water-skiing possible. But it also makes it possible for a vehicle to glide along a thin water film covering a smooth surface without its wheels rotating. The friction is extremely low. The fluid depth (water or slush) required for hydroplaning may be as low as 3 mm (1/10 inch) for smooth or worn tyres. In touchdown zones with rubber deposits, even thinner water films may reduce ground friction to zero.

The speed at which an aircraft starts hydroplaning depends mainly on the tyre pressure, rather than on the aircraft's weight (tyre footprint increases with increasing weight and constant tyre pressure). These speeds may vary from about 140 k/ph (around 70 kt) for a propeller transport aircraft to 190 k/ph (around 100 kt) for a jet transport and 300 k/ph (around 150 kt) for a heavy fighter.

Hydroplaning is especially dangerous in a crosswind when lack of ground friction may lead to loss of directional control (fig. 9.49).

Note that partial hydroplaning can be obtained at considerably lower speeds than those indicated in fig. 9.49.

When landing on wet runways it is advisable to use the minimum safe landing speed, to touch down with some rate of sink (not too smooth a flare) and to use spoiler, thrust reversers and brakes as early as possible. Asymmetrical use of brakes or reverse thrust may result in directional control problems on slippery runways.

The combined effects of tailwind, high touchdown speed, and aquaplaning may make the braking distance five times as long as the distance demonstrated in dry runway landing tests (fig. 9.50a). Fig. 9.50b shows the difference in braking distances on slippery runways when landing in tailwinds and headwinds.

Slick spots on runways may pose special problems for aircraft with anti-

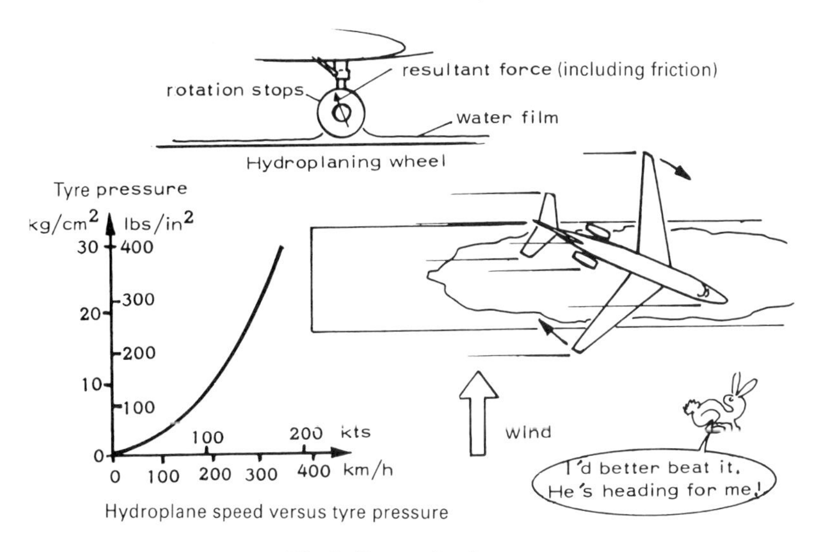

Fig. 9.49 *Hydroplaning*

Braking distances

(a) Effects of runway friction, tailwinds and touchdown speeds on braking

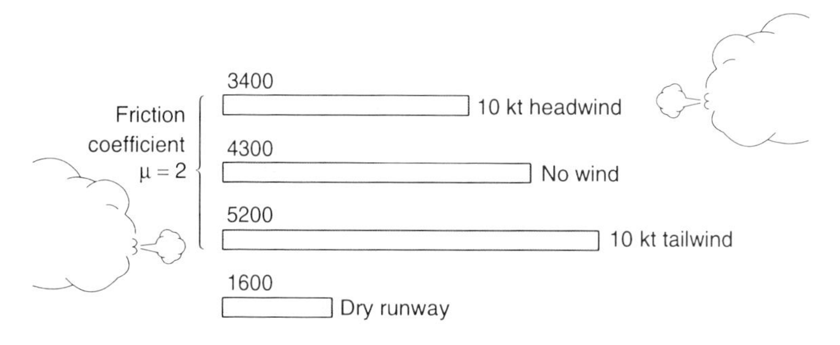

Braking distance in feet

(b) Effects of landing in tailwinds and headwinds on slippery runways

Fig. 9.50 *Factors affecting braking distances*

skid brakes. The brake system may not be able to react fast enough to the rapidly changing runway friction. As a result, the average braking action may be considerably lower than expected. Also, the wheels may lock when rolling on to slick spots with maximum braking. As a result, the tyres may fail as the skidding wheels run from slick spots to high-friction runway surface.

Aerodynamic braking is one way of reducing the speed of a swept-wing or delta wing aircraft during a ground-braking run when other methods fail (or when you want to reduce brake or tyre wear). The drag of these aircraft in a stalled or semi-stalled condition is so high that it serves as a very efficient means of braking to the speed where the angle of attack must be reduced due to insufficient pitch control (fig. 9.51). Note: aerodynamic braking in a heavy crosswind may result in loss of directional control, especially on slippery runways where asymmetrical braking may have little effect.

The smooth surface of a concrete runway may become very slippery with a millimetre-thin water film, especially if the surface was covered with dust after a long period of sunshine. Light rain mixing with the dust makes the surface oil-slippery. The problem can be eliminated by scoring the runway surfaces as undertaken at Stockholm's Arlanda airport (fig. 9.52 and fig. 9.53). The scoring made the wet-runway friction good at all speeds (fig. 9.53). It cost less than the landing-gear of a large transport aircraft.

Nobody has complained about increased tyre wear due to the scoring at Arlanda. Tyres do not wear fast on surfaces with constant friction. On

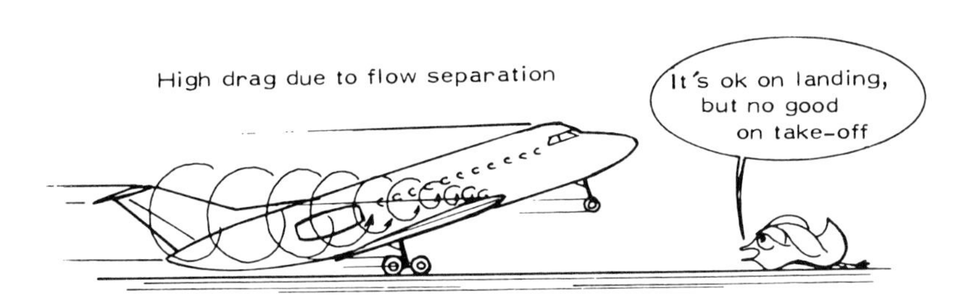

Fig. 9.51 *Aerodynamic braking*

surfaces with alternating good friction and slick spots, on the other hand, the wheels may lock on the slick spots and when the stopped wheels slide on to areas with good friction large pieces of tyre rubber may be ripped off.

In addition to the scoring, the airport management at Arlanda bought the best friction tester available (fig. 9.54). This tester is equipped with a

Fig. 9.52 *Runway surface scoring*

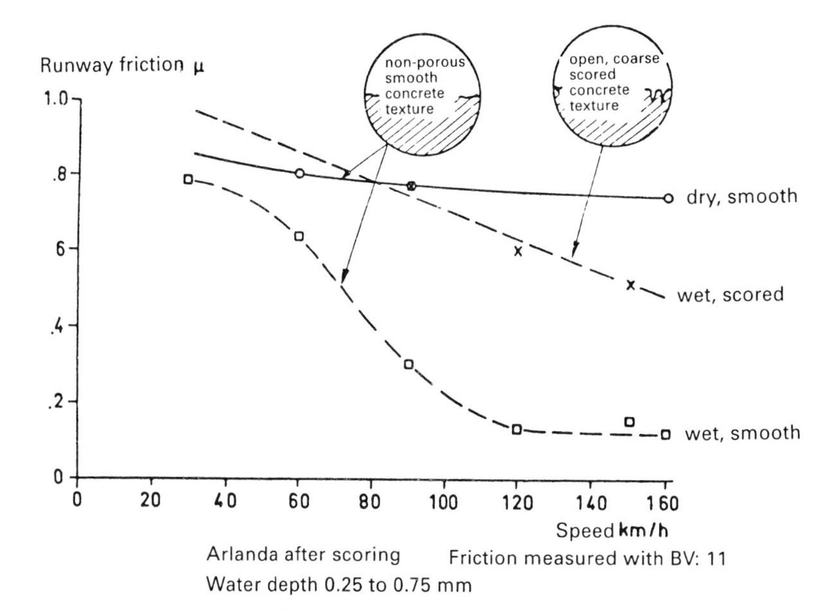

Fig. 9.53 *Friction coefficient of scored runway surface*

computer. Friction can be tested at low and very high speeds, and friction values can be reported to the tower while the tests are being made.

You might now assume that all friction testing and reporting problems at the airport had been solved. This, however, was not the case. One winter day with freezing rain, a Tupolev 134 landed. Deployment of the braking chute was delayed since 'braking action good' had been reported. The crew found that the runway friction was close to zero. They deployed the brake chute and 'stood on the brakes'. As a result, several main tyres exploded and the aircraft veered off the runway.

Fig. 9.54 *The* SAAB *friction tester*

The problem is that if the runway friction is measured shortly after freezing rain has started to fall, the friction tester tyre breaks through the thin ice and contacts the rough scored surface. 'Braking action good' will be recorded. However, a short time after the test, when the ice over the rough surface has become thicker, the tyres will no longer crack the ice and the runway friction will be very poor. Friction will deteriorate faster than the coefficients can be measured. Hence, in case of freezing rain, urea must be immediately spread on the runway surfaces. There is no time to wait for the usual slow deterioriation of friction.

Poor friction is not only a problem during wintertime. Runways with smooth surfaces may, as shown in fig. 9.53, become as slippery as ice-rinks with thin layers of water on them. This is especially true after long dry periods when the runway surfaces may be covered with layers of dust. The first light rain may make the dust-covered surfaces soap-slippery, especially when runways have rubber deposits on them.

Examples of the differences between theory, flight tests and practical operations are shown by the following two cases.

The flight manual of a subsonic fighter showed the braking distance after touchdown to be 500 m. On a beautiful day with the best runway friction, one of the test pilots managed to stop at this distance using maximum braking. Both main tyres blew! Not operational!

The runway requirement of a supersonic fighter design was 500 m. In order to meet this requirement, a unique aerodynamic high-lift design was developed. This design had, however, little effect on the landing distances compared to the distances required by earlier supersonic designs. The following was needed to meet the design requirement in landing:

- autothrottle to eliminate overspeed landings and the risk of dangerous speed losses during landing
- carrier-type landings without flares to reduce the touchdown scatter from ±300 m to ±50 m
- head-up display with good landing information
- drooped radar nose to make it possible to see over the nose to the touchdown point.

Together, these made repeated, short-field precision landings possible. Slippery runways are also discussed in chapter 10.

Chapter 10

Contaminated Wings, Contaminated Runways

Contaminated wings cause accidents. Contaminated runways cause accidents. Together they spell disaster.

The performance and handling qualities of all types of aircraft may be seriously degraded by skin contamination, wear, foreign-object damage and inferior maintenance. This is especially true for modern jet transports with wings optimised for high lift and low drag. However, it is surprising to see how large the effects of a little contamination can be on all aircraft in service today.

Surface contamination is not only caused by frost, snow and ice. Even insect contamination at the wing leading edges, mud spray and tropical rain showers affect performance. Leading edge wear caused by rain erosion, surface roughness caused by foreign-object strikes, and leakage into the airflow from pressure cabins and leading edge hot -air ducts caused by poor maintenance may also have considerable effects on aircraft drag, stalling speeds and handling qualities.

Since all aircraft take off and land with some margin to stall, pilots may experience that a little contamination and surface roughness has no effect because the aircraft took off and landed safely. This is a dangerous and misleading experience. There is always some degrading effect and, when it exceeds the stall margin, accidents happen.

It does not cost very much to be professional and maintain clean and smooth skin surfaces, and to be aware of the risks when in-flight contamination cannot be avoided. It may cost you your life to believe that a little frost has no effect and that the snow on the wings will blow off during take-off acceleration.

It is surprising that this is not obvious to everyone, as shown by the following stories.

Once upon a time in a northern territory there was a bush-pilot who said, 'Taking off with frost, snow or ice on the wings can be very dangerous unless you know what you do'. So he flew year after year with a little contamination on the wings thinking he was gaining experience and knew what he was doing. After ten years, he lost all of his margin to stall and was killed in a take-off accident. He now had gained the ultimate experience, useful to others but not to himself.

Recently, a pilot was loading his aircraft with newspapers in freezing rain. Just before taxying for take-off, snow began to fall. No de-icing was carried out. The aircraft stalled at 300 ft and dived into the ground. Dead pilot!

Some airport managers do not understand the need for proper runway care. Runways are rough, have holes in the surfaces, standing water, rubber deposits, slush, snow and ice, and do not have runway ends prepared to protect overrunning aircraft.

A pilot from a 'rough-runway country' landed by mistake on a runway with the asphalt removed to prepare for a new surface. When asked what he thought of the runway he said, 'Normal, normal'. All pilots are not spoiled.

During the early years of commercial aviation when passengers were flown in biplanes with struts and wires between the wings, the weight of the ice collected on fuselage, wings, wires and struts could force an aeroplane down. Today, when a light layer of dead insects along the wing leading edges can make an aeroplane stall before any stall warning has been received, weight is no problem compared to the aerodynamic effects of surface roughness.

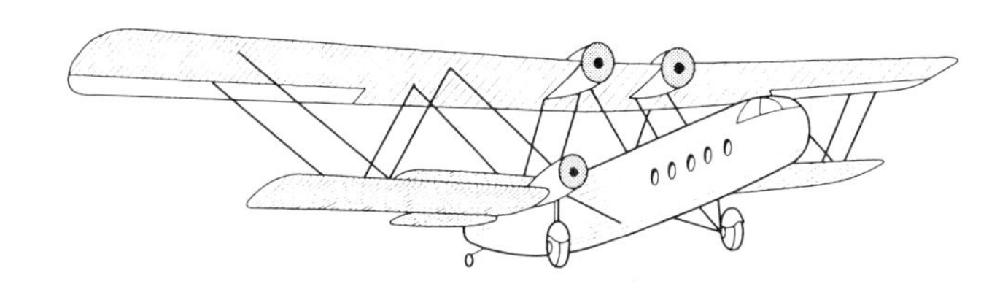

Fig. 10.1 *An ice collector (Handley Page HP42)*

When monoplanes were introduced the aerodynamic problems began to increase. However, many of these aeroplanes, such as the German Junkers Ju 52 with its corrugated skin, had fairly rough surfaces and thick boundary layers drowning many effects of surface contamination. When the DC-3 with its relatively smooth wing surfaces was introduced it became necessary to install pneumatic de-icing boots on the leading edges of the wings and tailplane in order to maintain acceptable performance during flights in icing conditions. This has unfortunately made many pilots believe that they

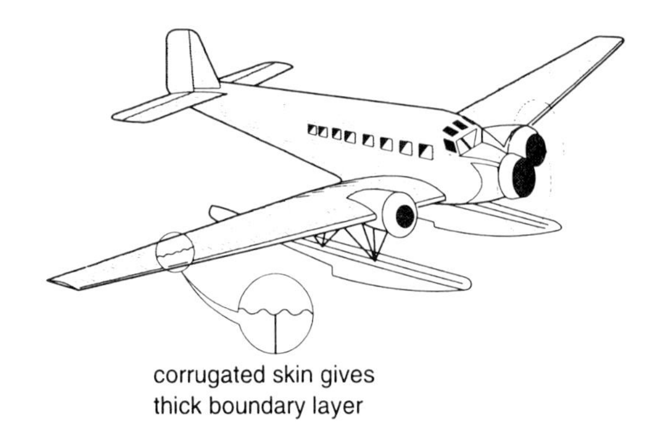

Fig. 10.2 *Junkers Ju 52*

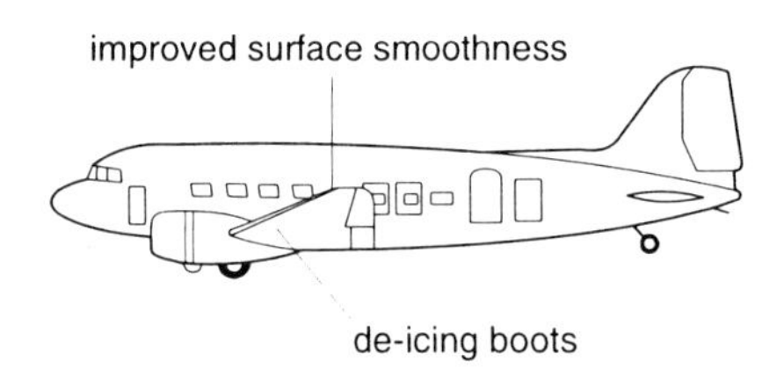

Fig. 10.3 *The DC-3*

are safe as long as the wing leading edges are kept clean. It has caused many accidents. The following is an example.

On a late, dark winter evening a DC-3 freighter landed for refuelling. It was snowing. When the captain, who remained in the cockpit, detected some activity on the left wing he opened the side-window and asked, 'Hey, what are you doing out there?' When told that the wing was being de-iced he shouted, 'Cut it out, I don't want it.'

So with the left wing fairly clean and a couple of inches of snow on the right, the crew taxied to the take-off runway. The aeroplane accelerated and rotated for take-off. When it did this, the left wing lifted and the right stalled. As a result, the aeroplane yawed 45° to the right, cut across the field and ran through a fence into the aeroplane scrapyard. It is still there.

Aeroplane development has continued. Aerodynamic research has made it possible to increase the maximum lift of wings by approximately 300% since the DC-3 era. This increases the contamination problems. The new, aerodynamically refined wings cannot take much of a boundary-layer disturbance without loss of lift and increase in drag. The better the wings are, the larger the losses can be. The performance of the super-aircraft may degenerate to the Ju 52 and DC-3 levels if the aircraft surfaces are not maintained smooth and clean.

Lift

The effects of flow disturbances on wing lift may be illustrated as follows. The boundary layer, caused by friction between the air and the wing surface, becomes unstable when it slows down and reaches a critical thickness. It separates from the surface and the wing stalls.

In front of the suction peak near the wing leading edge, the boundary layer is accelerated by the decreasing pressure, and no flow separation occurs. Behind the suction peak, however, the increasing pressure slows down the boundary layer and when it reaches a critically low speed and has become sufficiently thick it separates from the surface and the wing stalls (fig. 10.4).

The stall angle depends on the size of the suction peak and on the friction along the wing surface. For a given lift value, the suction peak can be reduced by cambering the wing profile and introducing extra camber to the chord by means of slats at the wing leading edge and flaps at the trailing edge. The camber spreads the suction peak over a larger portion of the chord. Thus the pressure gradient behind the peak, which the boundary layer has to flow against, is reduced and stall is delayed.

Since stall is affected by the thickness of the boundary layer it is obvious that the stall angle decreases with increasing friction. Modern wings are, therefore, made with the smoothest possible surfaces and the best possible camber.

Stall can be further delayed by re-energising the upper surface boundary layer with air accelerated from the lower wing surface through slots at slats

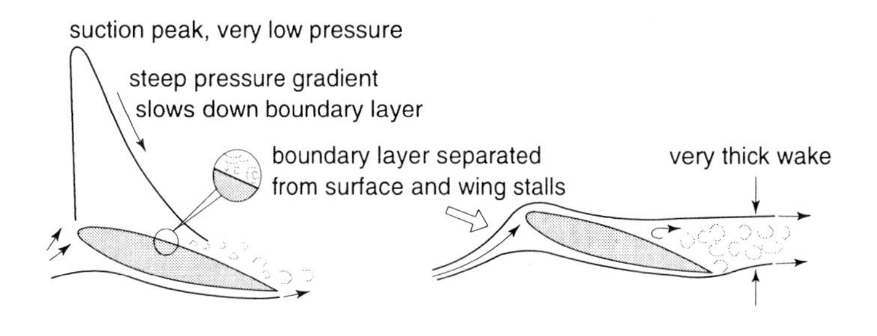

Fig. 10.4 *Stall*

and flaps. In this way it has been possible to obtain very high maximum lift coefficients reducing both take-off and landing distances.

A wing profile made of a thin, flat plate stalls at $C_L = 0.6$. The flow separates at the leading edge at very low angles of attack. However, a leading-edge vortex is created. The main flow jumps over the vortex and reattaches to the wing surface behind it. The vortex diameter increases with increasing angle of attack and the vortex finally breaks down at stall. A profile of this type can take considerable contamination without any serious effect.

Camber, slots and a smooth surface move the flow-separation point rearward on the profile and maximum lift increases. However, if the flow is disturbed by surface roughness there is always a risk that most of what has been gained through optimum design will be lost and the wing approaches flat plate performance. The potential losses are large (fig. 10.5).

There is no method to predict with any accuracy the lift losses caused by various types of boundary layer flow disturbances. In order to avoid surprises in the attempt to maintain certified stall performance it is necessary to maintain clean and smooth wing surfaces under all operating conditions and to add margins when this is not possible.

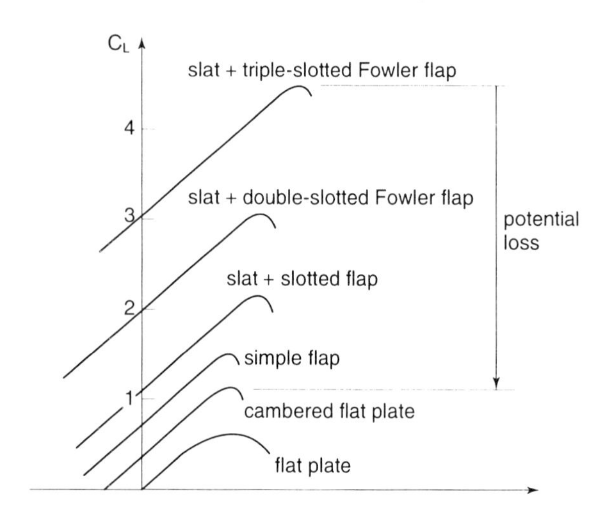

Fig. 10.5 *Potential lift losses*

Drag

A cylinder in frictionless flow has no drag. The reason for this is not only the lack of friction, but when there is no flow separation the flow velocities at the rear of the cylinder will be equal to the velocities at the front.

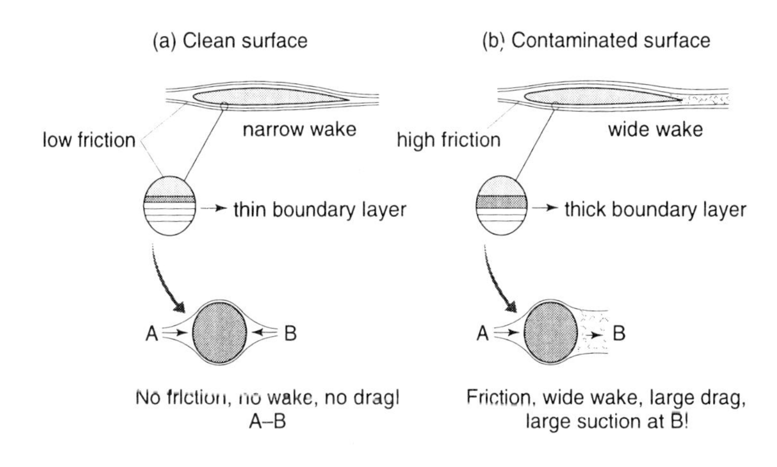

Fig. 10.6 *Friction drag and pressure drag*

At the front and rear stagnation points (A and B in fig. 10.6) the speed is zero. When the velocities are the same, the pressures against the cylinder surface are the same. The large rearward pressure force created on the front of the cylinder is, therefore, balanced by an equal, forward-acting force on the rear. In this case, 100% pressure recovery is obtained (fig. 10.6a).

In aircraft design, the goal will always be to come as close to 100% pressure recovery as possible. This can be done by combining the best possible streamlining with the manufacture of smooth surfaces.

In real flow with friction, the boundary layer will separate a short distance behind the pressure peaks at the cylinder and a very wide wake will be obtained (fig. 10.6b). Outside the wake the flow velocity will be high and the pressure low. As a result, a large suction force is obtained at the rear of the cylinder. In this case, the pressure recovery is very low and the total drag is dominated by the pressure drag.

If a streamlined profile is contaminated, the boundary layer thickness may increase rapidly and a very wide wake may be obtained. The pressure recovery at the rear of the profile will be poor and the drag force may approach the cylinder drag. This is especially true for low-drag profiles. The better the aircraft is, the larger the losses may be.

Low drag requires clean and smooth aircraft surfaces, including the fuselage. Fuselage flow disturbances may easily increase aircraft drag by 10% to 20%.

Handling Qualities

Good handling qualities at stall require good control in roll, pitch and yaw throughout the stall. The stall must be preceded by an easily recognisable natural or artificial stall warning. When the aircraft stalls it should nose down so slowly that the crew can recover before the aircraft enters a steep dive, preferably before the nose sinks below the horizon.

Good stalling characteristics require that the flow separation begins in the wing-root sections and spreads slowly outboard. This gives good roll-control and buffeting that may serve as a stall warning.

Small aircraft with straight wings can always be designed to have wing-root stall either by proper wing design or by installing stall strips on the wing leading edges in the root sections (fig. 10.7).

Swept-back wings are more difficult to handle. The boundary layer flows outboard when the angle of attack increases. As a result, the boundary layer thickness increases towards the wing-tips and when the sweep angle exceeds a certain fairly low value wing-tip stall is obtained (fig. 10.7). The boundary layer flow towards the wing-tips can be reduced by means of boundary layer fences which drain off part of the low-energy air in the chordwise direction. This delays but does not prevent wing-tip stall.

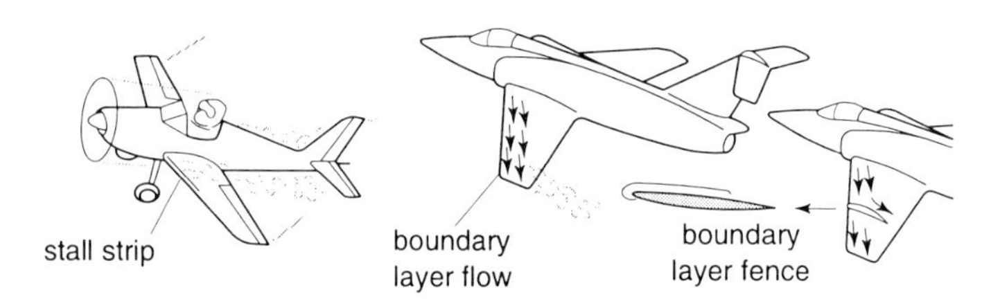

Fig. 10.7 *Wing-root and wing-tip stall*

When the wing-tips stall, the wing lift resultant moves forward. This gives the aircraft a nose-up moment. If the resulting pitch-up becomes too fast, the aircraft may rotate into a complete stall before the crew can prevent it. For aircraft with body-mounted engines and long forebodies, this may result in rotation into deep stall. In this condition the fuselage vortices create a large downwash on the tailplane locking the aircraft in a high angle of attack attitude from which recovery is nearly impossible due to insufficient pitch control (fig. 10.8). For this reason, wing-tip stall must develop slowly in order to make the pitch-up controllable. Roll control devices (spoilers and ailerons) must be located inboard of the tip-stall region to prevent loss of roll control at stall entry.

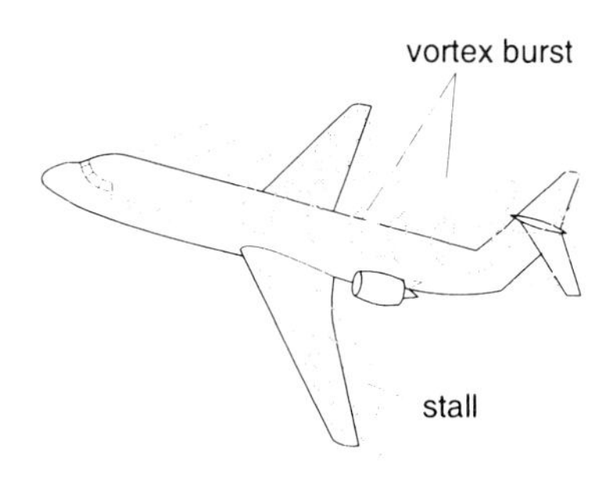

Fig. 10.8 *Deep stall*

Careful design is required to obtain good stalling characteristics. One example of such a design is shown in fig. 10.9. A small stall strip has been placed immediately inboard of a boundary layer fence on the leading edge. The strip is roughly 2.5 inches wide and protrudes ⅛ inch through the leading edge. The stall strip reduces the local stall angle and as a result the aircraft, which has a moderately swept-back wing, has a wing-root stall giving a slow nose-down stalling motion which is easy to control.

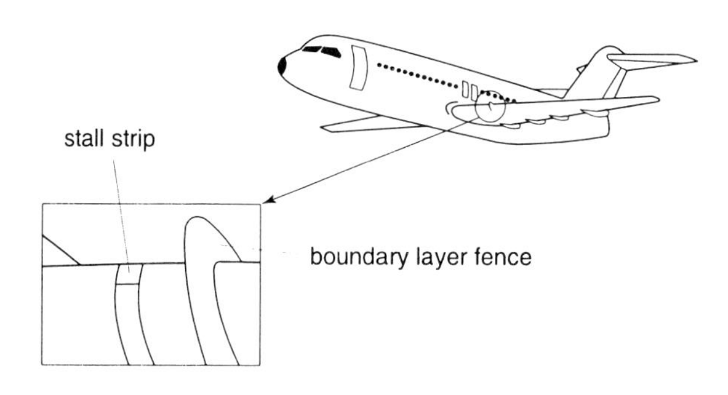

Fig. 10.9 *Wing-root stall strip*

The good handling qualities in stall of this aircraft can be completely destroyed by boundary layer flow disturbances. The positive effect of the small stall-strip is easily overwhelmed by frost, snow or ice on the wings. This is a problem for all aircraft. Contamination does not only have detrimental effects on lift and drag. Even the handling qualities of the aircraft may change dramatically.

Tail Stall

Good pitch control at take-off and landing is an obvious flight safety requirement. This requirement cannot be met if the tailplane stalls at normal approach and landing speeds.

The tailplane balances the pitching moment of the wing and fuselage. When the aircraft's centre of gravity is located forward in the approved c.g. range the moment is negative at all speeds and a tailplane down-load is always required for trim. Theoretically it can be shown that the tailplane lift coefficient is roughly proportional to the pitching moment of the wing and fuselage coefficient.

When trailing edge flaps are extended, a rearward shift of lift is obtained and the wing and fuselage nose-down pitching moment increases (fig. 10.10). Hence, the down-load on the tailplane increases. If there is a risk of tailplane stall this will, therefore, increase with increasing flap angles.

Assuming that the aircraft's c.g. is located at the centre of pressure (c.p.) of the wing and fuselage, close to 25% of the wing's mean aerodynamic chord (m.a.c.), the changes in wing lift due to changes in angle of attack have no moment around the c.g. Both pitching moment coefficient and tailplane lift coefficient remain constant, independent of speed or load factor.

If the aircraft is equipped with a fixed tailplane and trim is obtained only by means of the elevator, the angle of attack of the airflow at the tailplane leading edge is low at low speeds since the downwash behind the wing is more or less cancelled by the high angle of attack of the aircraft. The main trim load on the tailplane is, therefore, located aft on the tailplane, near the upward-angled elevator.

When the elevator trim angle decreases with increasing speed, the trim load on the tailplane moves forward. At high speeds, when the elevator angle is negative, the leading edge loading may come close to stall. Therefore, even if the loading of the fixed tailplane is constant at a forward-located c.g. the critical condition for tailplane stall increases with increasing speed.

This problem limits the maximum speed at which flaps can be lowered. Flap extension above the maximum flap speed may result in tailplane stall and an uncontrollable dive. Accidents have been caused by flaps being lowered at high speed.

If the pitching moment is trimmed by means of an all-moving tailplane powered by means of a trailing edge trim surface, the problem changes completely. In this case, the maximum down-load at the tailplane leading edge is obtained at minimum speed when the trim surface is angled downward and an s-shaped, chordwise lift distribution, with the down-load at the forward area of the tailplane is obtained. When the speed is increased and the trimming surface moves upward, the down-load on the tailplane moves rearward and the leading edge area is off-loaded. In this case, the risk of tailplane stall increases with decreasing speed and increasing load factor. This shows that the risk of tailplane stall depends very much on whether the pitching moment of the wing and fuselage is trimmed by means of an elevator or by trimming with the whole tailplane. In the first case, the risk of tailplane stall increases with increasing speed and, in the second, the risk is highest at minimum speed.

When the c.g. moves forward, the lift of the wing and fuselage acting behind the c.g. gives an additional nose-down pitching moment that increases as the wing lift coefficient increases, i.e. with decreasing speed and increasing load factor. Hence the speed at which tailplane stall may be obtained with a fixed tailplane decreases as the c.g. shifts forward. The risk of tailplane stall is highest when the landing flaps are extended, the c.g. is at the most forward position and the tailplane is trimmed for maximum nose-down.

For well-designed aircraft the wing and fuselage pitching-moment changes obtained when flaps are lowered or retracted are balanced by downwash changes at the tailplane giving automatic changes of tailplane trim load. Very small pitch disturbances are, therefore, obtained during flap movements. This may give a false impression of safety, an impression that the load changes at the tailplane are small. For large flap angle they are not small, especially if the flaps are very efficient and move rearward during extension. This gives large pitching-moment changes.

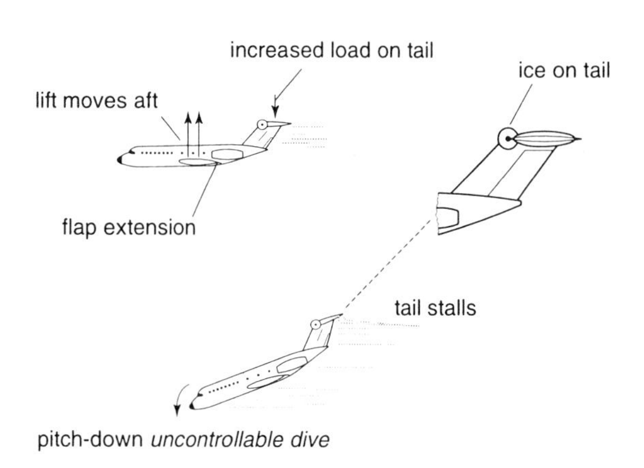

Fig. 10.10 *Tailplane stall*

Ice on the tailplane leading edge reduces tailplane lift and increases the risk of tailplane stall. Partial flow separations at the tailplane, obtained when the tailplane is loaded to a near-stall condition when flaps are lowered, reduce tailplane damping efficiency. This may result in unpleasant pitching motions of the aircraft. Also, flow separations at the tailplane reduce the aerodynamic balancing of the elevator and elevator control forces may increase considerably. The flow separations may also shake the elevator. This may vibrate the control column. Pitching motions, increased stick forces and control column vibrations are signs of imminent tailplane stall and the flaps should be retracted. If the tailplane stalls, the aircraft may pitch down violently and enter a dive or bunt from which recovery is not possible unless the flaps are immediately retracted.

The risk of tailplane stall on most aircraft with elevator control increases with increasing speed. In icing conditions the crew may be faced with wing stall if the speed is reduced and tailplane stall if it is increased.

Directional control in the case of single engine failure at low speeds for multi-engine aircraft can be a problem at speeds well above minimum control speeds. Some aircraft roll on to their backs unless the crew reacts quickly and knows exactly what to do. During take-offs and landings in icing conditions, ice on the leading edge of the fin may make directional control impossible if an engine fails.

Certification tests of performance, stability and control are made with new, clean aircraft. When the boundary layer flow along the surfaces of these aircraft is disturbed by contamination or other factors affecting the flow, a dramatic degrading of performance and handling qualities may be expected, especially with aircraft with very advanced aerodynamic designs.

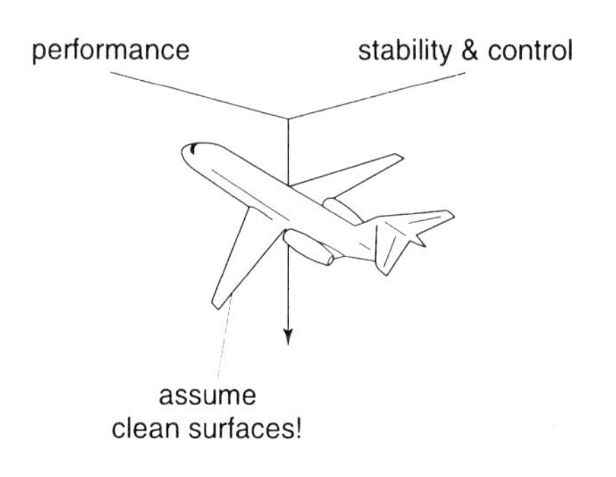

Fig. 10.11 *Handbook data*

Practical Experience

Theoretical discussions may be educational but practical experience gives a more convincing impression of what may happen if an aircraft's surfaces are not maintained clean and smooth. In the following, flight test results and accident data are used to illustrate serious effects of disturbances of boundary layer flow.

The aircraft mentioned are some of the best produced and are affected by contamination to the same extent as any other aircraft flying today. They are mentioned simply because information was available to the author.

Flight Test Data

Saab MFI-15

In order to investigate the effect of frost on the wing, the aircraft was left outside on a winter night. In the morning the wing was covered with hoar frost. Take-off tests showed that the stalling speed had increased by 30%. This corresponds to a roughly 50% reduction of the maximum lift coefficient.

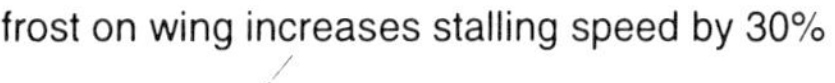

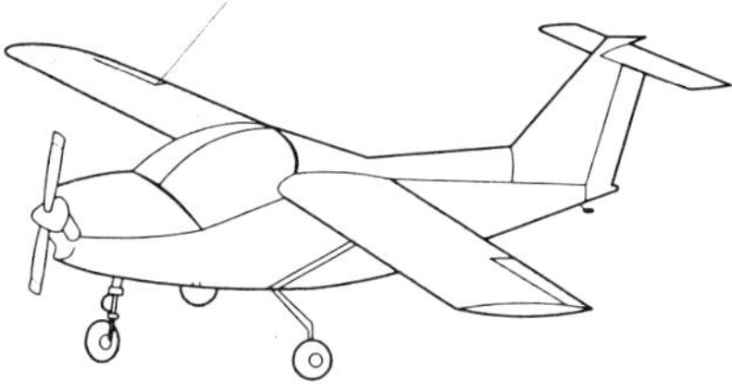

Fig. 10.12 *Effect of frost on a MFI-15*

The MFI-15 was later equipped with a research wing having a maximum lift coefficient of 4.5 with extended Fowler flaps but no slats. Flight tests showed that the stalling speed of this aircraft increased by 10% in a light drizzle. Be aware of this problem when you land any modern transport aircraft in a tropical rainstorm.

Cessna

Flight tests of a single-engine Cessna showed that ice on the wing leading edges rapidly increased the stalling speed. Six millimetres of ice (¼ inch)

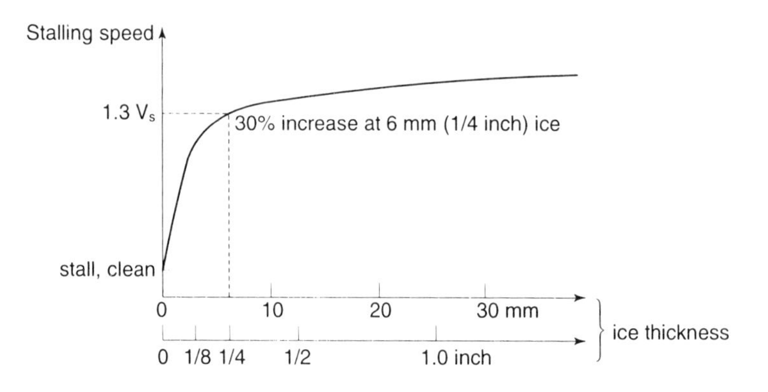

Fig. 10.13 *Effect of leading edge ice on stalling speed*

resulted in a 30% increase in the stalling speed (fig. 10.13). Further ice build-up had little effect on the stalling speed but the drag increased. The test shows that performance deteriorates as soon as ice is formed and that stalling speeds increase rapidly with increasing thickness of ice.

Aircraft with De-icing Boots

Pilots flying aircraft with pneumatic de-icing boots know that stalling speeds increase until the ice is broken and also when the boots are inflated. However, flight tests have shown that bits of ice may cling to the rubber boots after the ice has been broken. Together with ice remaining on other parts of the aircraft after in-flight de-icing, this affects stalling speed and drag. Count on at least a 10% increase in the stalling speed and increases in drag.

Pneumatic de-icing can only remove ice formed on the boots. Ice formed behind the boots cannot be removed. This has caused many accidents. For this reason do not extend flaps at speeds so high that the forward stagnation point will move to the upper surface of the wing and ice forms behind the rear of the boots and do not fly at speeds so low that ice may form behind the rear of the boots on the wing's lower surface.

At temperatures near freezing, runback 'glass-ice' may form on the wing leading edges. Since the water at these temperatures flows some distance before it freezes, the glass-ice gloves growing on the leading edges may extend behind the de-icing boots before they are broken. When the boots are activated, sharp ice ridges are, therefore, formed behind the boots (fig. 10.14).

As the flight continues, an increasing amount of ice may grow on the ridges. This ice cannot be removed by the boots. Finally, the effects on stalling speed and drag become so large that climb or even continued level flight becomes impossible. A fatal accident can only be avoided by an

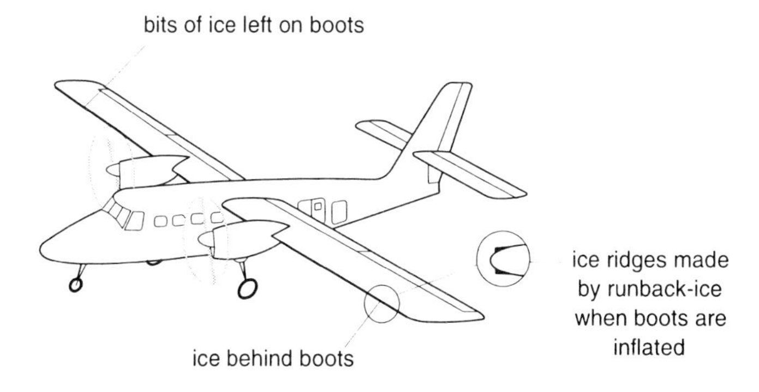

Fig. 10.14 *Problems with pneumatic de-icing*

immediate high-speed descent to a flight level with warmer air as soon as the problem is discovered.

This problem concerns all aircraft using pneumatic de-icing. Breaking the fairly smooth glass-ice gloves may do more harm than leaving them on the leading edges.

DC-9

During a post-maintenance flight test the aircraft was to be flown down to stick-shaker speed with slats and flaps retracted. Before any stall warning had been obtained, the aircraft rolled on to its back. Repeated tests gave the same result. An aircraft check after landing showed that the wing leading edges were contaminated with layers of dead insects. The wings were wiped clean and the test was repeated. The stick-shaker speed was now reached with full roll control.

F28

During a flight test down to stick-shaker speed after repairs to a fire-damaged area, the aircraft rolled on its back before the stick-shaker was activated. Repeated tests gave the same result.

A check after the flight showed that air was leaking from a wing leading edge hot-air duct, past the edges of poorly sealed inspection strips into the airflow. At high angles of attack this triggered flow separation on one wing resulting in an uncontrollable roll.

The flight tests of the DC-9 and the F28 show that jet transport performance can be seriously affected by relatively small flow disturbances of the leading edge boundary layer.

This problem was also experienced with a swept-wing jet fighter. After a few years in service, roll disturbances began to develop during normal landing approaches. A careful study of the problem showed that the roll disturbances were the result of small flow disturbances caused by rain erosion of the leading edge. Plastic tape on the leading edges solved the problem.

These tests show that it is advisable to keep the leading edges of the wing and tailplane as clean and smooth as possible. All types of contamination, rain erosion, dents in the skin plates, poorly made skin repairs, and leakage from hot-air ducts may have considerable effects on roll stability and stalling speeds.

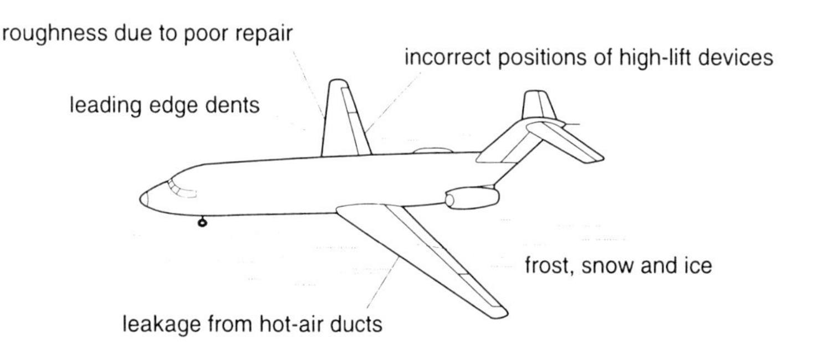

Fig. 10.15 *Factors affecting roll stability and stalling speeds*

Boeing 737

Reports had been received that the performance of the aircraft could be seriously affected by frost, snow and ice contamination on the wing's upper surfaces. For this reason, tests were made with simulated frost on the wings. These tests showed that in the worst case the following could happen:

- stalling speed increased by 20% to 28% depending on the flap angle; the largest effect was obtained with take-off flaps
- the aircraft could stall before the stick-shaker was actuated
- the aircraft could stall during take-off rotation unless the rotation was made slowly or the take-off speed was increased by a few knots
- the aircraft pitched nose-up during lift-off due to wing-tip stall
- single-engine climb became impossible with frost on the wings.

The simulated frost used in the tests was not as 'hairy' as real hoar frost. For this reason the effects of real frost are probably larger than the effects shown by the tests.

By making these tests, Boeing has done much to show the risks of taking off with frost on the wings. The problems revealed do not only concern the 737. All modern jet transports are probably similarly affected.

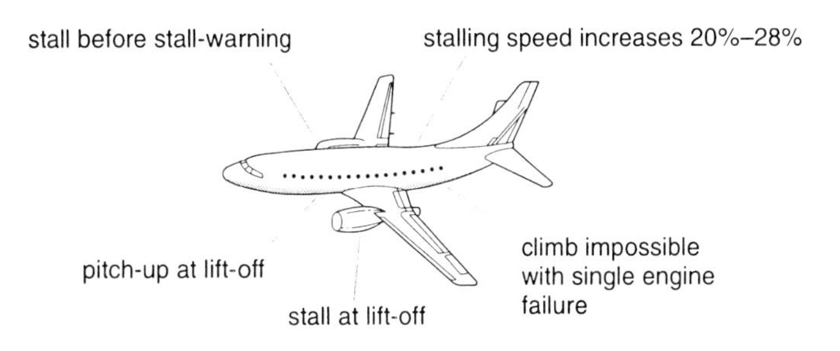

Fig. 10.16 *Risks during take-off with frost on the wings*

Frost on the tailplane and fin reduces longitudinal and directional stability and control and must, therefore, be removed. Frost on the fuselage may have considerable effect on fuselage drag. The effect of a flow disturbance of the fuselage boundary layer was well demonstrated by a leakage from the pressure cabin through a door seal into the outside flow. At cruising speed, it increased aircraft drag by 10%. Don't accept fuel-money wastage by tolerating fuselage flow disturbances!

Incidents and Accidents

- A Cessna 176 had been parked outside in a winter climate for a few days. When the pilot, an experienced military pilot, came to take the aircraft for a joyride he wiped frost off the windshield but not from the wings.

  The take-off was made from a grass strip. Acceleration to rotation speed was not slow enough to alert the pilot. However, after lift-off at full power, the aircraft refused to climb. It floated in the ground effect and finally settled on a field a couple of hundred metres from the end of the grass strip.

  A military pilot taking off in a high-powered delta wing fighter does not have to worry about wing frost but he sure has to learn to keep wings clean when he flies low-powered general aviation aircraft.

- The airtaxi Beech Baron stood parked for a few hours while the pilot waited for his passengers. Snow was falling. When the passengers arrived somewhat earlier than expected the pilot hurriedly wiped the snow off the wings and scraped ice off the leading edges. He did not take time to remove ice patches from the upper surfaces of the wing.

The take-off acceleration at first appeared to be normal but when the speed increased and the drag-effects of the ice patches began to be felt, the acceleration rate decreased markedly. Lift-off was made at the end of the 900 m long airstrip and the aircraft knocked down approach lights during the shallow climb. Fuselage skin plates were damaged but fortunately there were no hills preventing climb to a safe altitude.

After this incident a large number of pilots were asked if they believed that it was sufficient to clean the wing leading edges in order to maintain the aircraft's performance. Nearly 100% believed this was the case, a sign of very poor pilot education.

- The Aero Commander had been parked outside overnight and the wings were covered with frost. The pilots taxied to the runway without de-icing the aircraft.

  Acceleration to lift-off was made with good margin to the runway end. However, after rotation the aircraft refused to climb. It flew in ground effect along the runway, passed the airport fence, came over down-sloping ground where it lost altitude and crashed into the wall of a factory.

- A Boeing 737 rolled 45° to the left after take-off. The crew checked the roll and found that the rolling moment disappeared when the flaps were retracted. During taxying for take-off the crew had used the reversers for braking because the taxiway surface was very slippery. Light snow had been blown up by the reversers. It had melted and been deposited as ice on the left-side flaps (fig. 10.17).

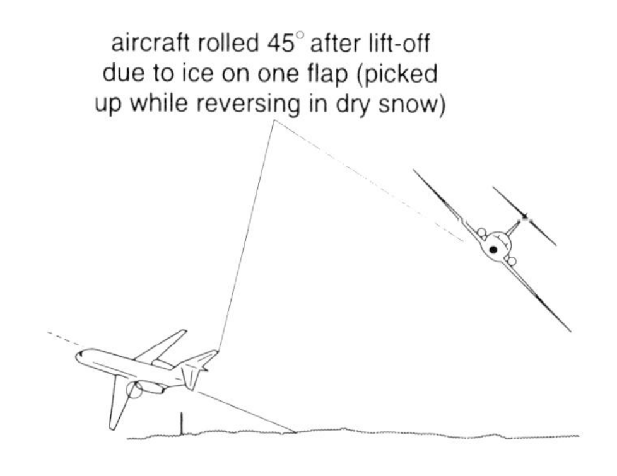

Fig. 10.17 *Melted snow on flaps*

- During a snowfall a Boeing 737 stood at the runway ready for take-off. The captain had two line-mechanics sweeping the snow off the wings. When they were clean the aircraft began the take-off roll. At lift-off rotation the aircraft rolled to the right. The captain quickly lowered the nose and continued the acceleration to a higher take-off speed. When the aircraft was checked after a precautionary landing it was found that the mechanic sweeping the right wing had walked backwards while sweeping and had left compressed snow footprints on the wing. The left wing was clean.
- An F28 stood parked with one wing outside the hangar on a cold night. In the morning the wing was covered by a layer of frost. The wing was not de-iced before the aircraft entered service. After lift-off, it rolled on to its back and crashed (fig. 10.18).
- A DC-9 continued to pitch up after lift-off on a dark winter evening. Full forward control movement was required to bring the nose down to normal attitude. Wing checks after landing showed that small patches of ice had not been removed from the upper surfaces of the outboard wing sections (fig. 10.19).

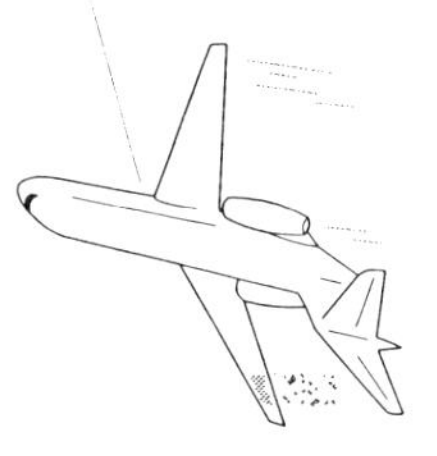

Fig. 10.18 *Take-off with frost on one wing*

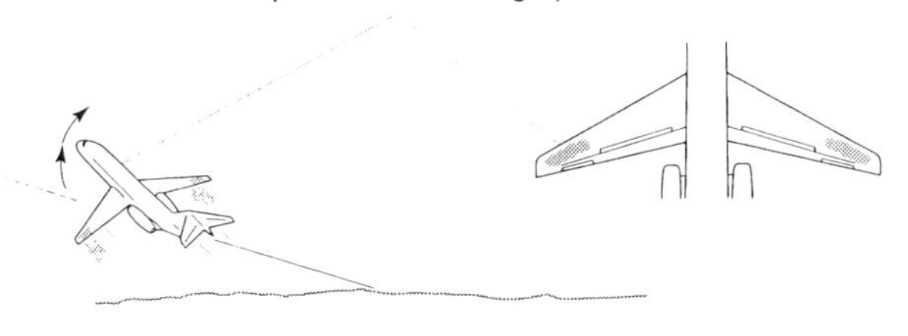

Fig. 10.19 *Ice patches near the wing-tips*

Lack of understanding or awareness of the risks of taking off with contaminated wings cause a number of aircraft crashes every year. Quite often, large transport aircraft are involved. In some of the cases the aircraft have been de-iced before leaving the gates but the time between de-icing and take-off has been too long, resulting in dilution of the de-icing fluid. Snow falling during taxying melted and froze on the wings. Some of the fairly recent cases are listed below.

- A Boeing 737 which crashed during take-off from Washington National Airport had been de-iced but the time between de-icing and take-off was too long. A contributing factor to the accident was ice on the total pressure heads in the engine air intakes resulting in too low a power setting being selected by the pilots during take-off.
- A DC-9 taking off from Denver, Colorado had also been de-iced but the time between de-icing and take-off was too long. After lift-off the aircraft rolled and crashed.
- A DC-8 taking off from Newfoundland had not been de-iced. During take-off in icing conditions it lost height and crashed.
- An F28 crashing during take-off in snow, Dryden, Ontario had not been de-iced/anti-iced and the snow had not been swept off the wings before take-off. It did not climb after lift-off and collided with trees beyond the end of the runway.
- An F28 which crashed during take-off from La Guardia, New York had been de-iced but the time between de-icing and take-off was too long. After lift-off the aircraft rolled and one wing struck the runway surface. The aircraft slid to a stop in the river beyond the end of the runway.

If we take all of these crashes and add all lesser-known accidents happening to regional and general aviation aircraft and to aeroplanes in countries which do not report accidents we find that billions of dollars are lost each year due to lack of understanding or care of contamination problems. An enormous cost for accidents that can be prevented by keeping aircraft clean!

Tailplane Ice

Ice on the tailplane leading edges has caused many incidents and accidents. A few cases are described here.

- The crew of a DHC Twin Otter extended landing flaps after flying in icing conditions. The aircraft immediately went into an uncon-

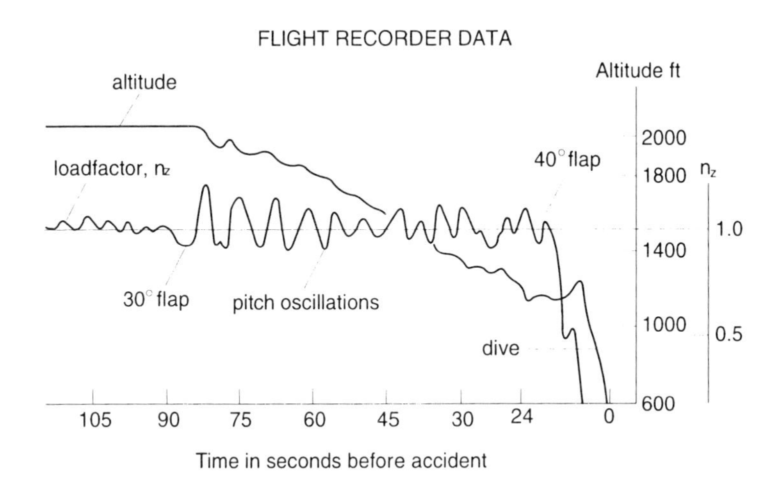

Fig. 10.20 *F.D.R. data, tailplane stall of Vickers Viscount*

trollable bunt to inverted flight. The crew retracted the flaps and rolled the aircraft back to an upright position.

- A Vickers Viscount was on landing approach. When 30° approach flaps were selected the aircraft began to oscillate in pitch. The flight continued along the approach path with slowly decreasing oscillations until the captain asked, 'Do you want 40° flaps now?' 'Yes,' answered the first officer who was flying. 'O.K., selected moving and you have it,' said the captain. At that moment the aircraft pitched down violently and dived into the ground (*see* flight recorder data in fig. 10.20).

The accident was caused by ice on the tailplane's leading edge. Flight tests after the accident showed that ice could form on the tailplane even when the anti-ice system was switched on.

Review of accident data showed that several similar accidents had happened. However, there was no information in the flight manual concerning the risk of tailplane stall. A major European airline prohibited the use of flaps during icing conditions. However, the pilots did not remember the reason for the restriction.

The pitching oscillations obtained after selection of flaps in fig. 10.20 are a result of reduced longitudinal stability due to ice on the tailplane's leading edge.

The climb recorded immediately before the dive is a false indication caused by airflow across the fuselage nose obtained when the aircraft pitched down. This cross-flow reduced the pressure at the static ports located on both sides of the nose. As a result a climb was recorded when the aircraft actually pitched down into a dive. The false climb indication shows that the rate of pitch-down must have been very high. During slow pitch-down motions no cross-flow is obtained.

- A BAe-31 Jetstream on approach to an airport in icing conditions came in too high and probably picked up considerable speed during the crew's attempt to descend to the correct altitude before reaching the airport. Approximately 400 feet from the runway the aircraft suddenly pitched down and dived into the ground. The flap angle at impact was 50°. The probable cause of the accident was tailplane stall due to ice and high speed. The crew may have intentionally kept the speed high in an attempt to avoid wing stall without realising that this could result in tailplane stall.

  Flight tests of the Jetstream after a similar accident showed that with ice on the tailplane's leading edge the aircraft pitched down when 50° flap was selected without working de-icing boots. Icing test regulations did not account for possible tail stall with ice on the tailplane's leading edge at this condition and the aircraft had not been tested to determine tailplane stall speed with tailplane ice and landing flaps down.

- A DC-9 on approach in icing conditions had the wing anti-ice system on. Tail anti-ice was off since both could not be selected at the same time.

  When approach flaps were selected the aircraft pitched down and abnormal pull in the controls was required to prevent a dive. The crew immediately switched to tail anti-ice. After a while, when the ice left the tailplane, the aircraft suddenly pitched up and a quick forward control movement was required to prevent a pull-up.

Some aircraft have tailplanes with inverted camber in order to increase the maximum negative lift of the tailplane. This does not mean that tailplane stall cannot occur since the potential lift losses due to ice on the leading edge increase with increasing maximum lift coefficient. Be sure that the tail anti-ice system works on aircraft of this type and do not extend flaps unless you are sure that the tailplane's leading edges are free of ice.

Other aircraft are not equipped with tailplane anti-ice systems. This is only safe if the maximum down load on the tail is lower than the load that can be supported with ice on the tailplane's leading edges. Aircraft designed to fly in icing conditions without a tailplane anti-ice system must still have the tailplane de-iced and anti-iced before take-off. Tailplane contamination decreases longitudinal stability and control. In a critical situation this may contribute to an accident. In severe icing, unusual pitch disturbances should be watched for and flaps retracted if uncommanded pitch-downs occur.

Critical Icing with De-icing Boots

A Mitsubishi MU-2 was cruising on autopilot at high altitude in icing conditions when it rapidly lost speed and began to shudder in pre-stall buffeting. The crew managed to regain control by diving down to a flight level with warmer air.

The probable cause of this incident was that runback ice ridges were formed behind the wing de-icing boots when the boots broke the ice glove. Ice build-up on the ridges during continued flight combined with ice formed on the rest of the aircraft resulted in large increases in drag and stalling speed. Ice build-up behind de-icing boots has caused many fatal accidents.

Flight Tests in Icing Conditions

All transport aircraft discussed have been flight tested in icing conditions and are certificated for flight in icing conditions. In spite of this, accidents happen.

One problem with flight tests in icing conditions is that it is difficult to know if the tests have been made in the most severe conditions. Large differences in performance and handling qualities may be obtained with flight tests made in inland climates compared to coastal climates.

Artificial ice mounted on leading edges and simulated frost on wing surfaces may not have the same effects as natural ragged ice and hairy frost. Flight tests using wooden 'horn-ice' on the tailplane gives less lift reduction than rough sandpaper on the leading edges.

Approval for flight in 'all icing conditions' does not mean that the aircraft is problem-free in severe icing. All aircraft, even the best, may run into problems. The operators must know what these problems are and the crews must act to avoid them and stay away from severe icing conditions.

Aircraft Without De-icing Equipment

Aircraft which are not designed for flight in icing conditions should never be flown into known icing weather. The combination of increased drag due to airframe ice and decreased thrust due to propeller ice may make continued level flight impossible.

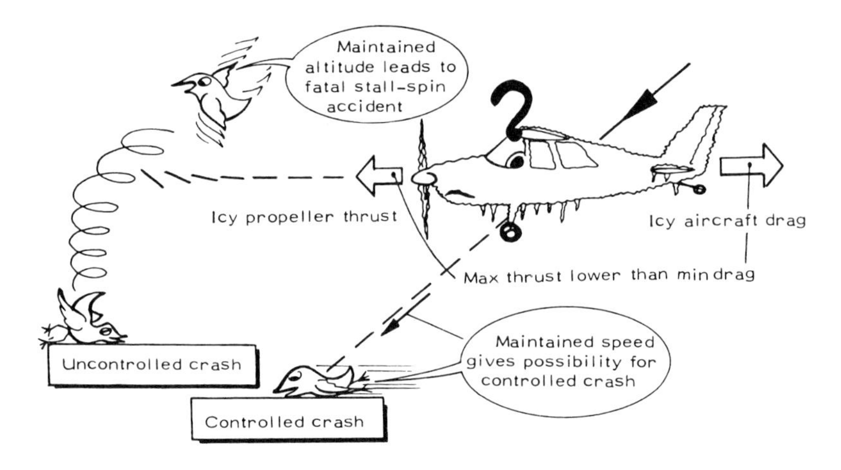

Fig. 10.21 *Over-iced*

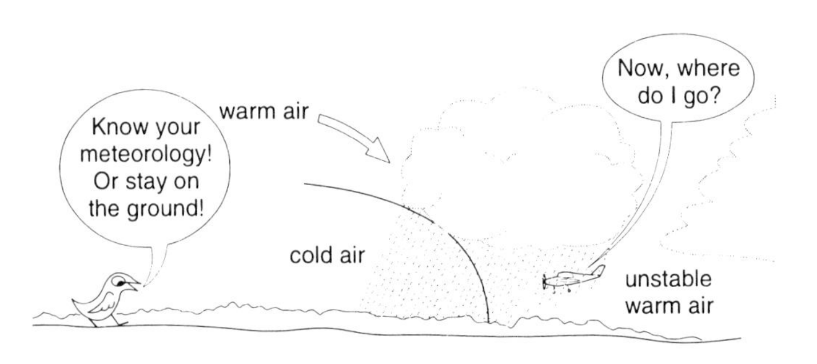

Fig. 10.22 *Typical icing conditions*

Pilots should remember that even light icing may increase stalling speeds by 30% and that attempts to maintain altitude when speed is lost may lead to a stall–spin accident. In that case, it is safer to maintain speed and make a controlled crash-landing if conditions do not improve before ground level is reached (fig. 10.21).

Since drag increases and thrust decreases the longer a flight in icing conditions continues, it is obviously necessary to try to get out of the icing weather as soon as possible. Pilots flying aircraft which are not designed for flight in icing conditions should, therefore, never take the risk of running into icing weather unless they know enough meteorology to avoid the icing clouds, or climb or descend to a safe flight level as soon as ice begins to form on the aircraft. Keep in mind that it may be very difficult for meteorologists to predict the icing risks accurately. Ice may form in one cloud but not in the one a short distance away. Light icing reported by a jet transport may be severe for a small general aviation aircraft.

During approach for landing through a layer of icing clouds with an aircraft not equipped with good de-icing equipment, the following should be observed.

- Never accept a shallow approach from A.T.C. Go through the icing altitudes at the highest safe rate of descent.
- Maintain an approach speed considerably higher than the normal speed since the stalling speed probably has increased by 30%.
- Watch out for tailplane stall when selecting approach and landing flaps. Retract flaps if unusual pitching oscillations or a nose-down pitching motion occur.

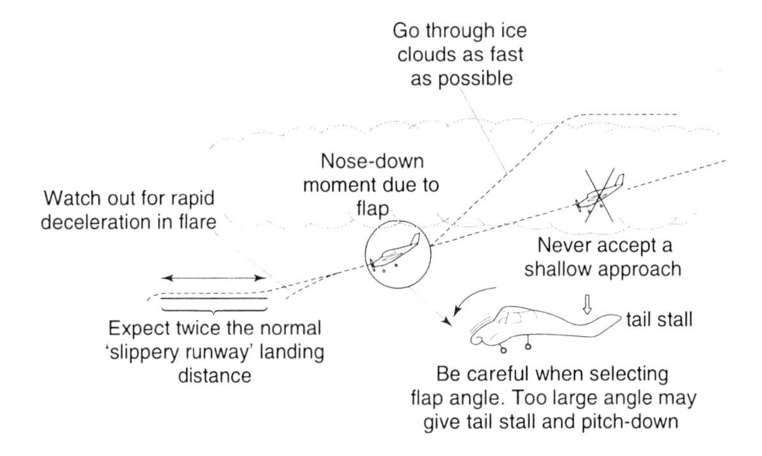

Fig. 10.23 *Landing approach through icing clouds*

- Watch out for rapid loss of speed during the landing flare due to increased drag from flow separation when the angle of attack is increased.
- Count on needing at least twice the normal landing distance due to increased stalling speed and on the possibility of a slippery runway surface.

The following accident shows what can happen when the awareness of the risk of flying in icing conditions is unacceptably low among aviation personnel.

The twin-engine Partenavia was not approved for flight in icing conditions. It had an electrical anti-ice system for the propeller but no wing or tail de-icing boots and no windshield anti-ice heating.

On the day of the flight the aircraft was released from maintenance with only the left propeller's anti-icing system working. The weather was poor and all airports west of the departure point were closed due to low visibility and extreme icing conditions. The pilot accepted the aircraft as released and took off for a one hour mail flight to the east. During turnaround at the destination airport he told the ground crew that he had run into considerable icing during the flight.

After one hour on the ground the pilot took off into the slowly deteriorating weather for a homebound mail flight. Roughly halfway home he contacted an international airport north of his flight path asking for permission to land due to severe icing. The request was granted and he began his descent. He informed A.T.C. that he could not maintain level flight but did not declare an emergency.

Since traffic was low and the weather was nearly calm, the landing approach could have been made from a southerly direction with plenty of altitude to spare. However, A.T.C. sent him north to a fix for approach to a southbound landing.

During the approach the aircraft collided with a tall tree and crashed in a forest a couple of kilometres from the runway. Fire broke out and the pilot was killed. Approximately one wingspan to the right of the tall tree, the forest was low and had the pilot flown here he would have reached an open field before he descended into the ground. The front windshield was probably covered by ice restricting his forward view. Visibility was reasonably good in the crash area.

In the above case the following contributed to the accident.

- The maintenance crew by releasing an aircraft with only one propeller anti-ice system working in winter conditions.
- The pilot by accepting the aircraft and the flight into known icing conditions.
- The pilot by not declaring an emergency and by not protesting against the long approach given by A.T.C. when he could not maintain altitude.
- A.T.C. by not clearing the pilot for a straight-in approach from the south when they were informed that he could not maintain altitude, even if he did not declare an emergency. A straight-in approach would have prevented the accident.

Leading Edge Ice Formation

Under certain atmospheric conditions, drops of water striking the leading edges of the wings and tail surface (and other parts of the aircraft) freeze. Depending on the temperature the water freezes on impact or flows some distance before it turns to ice. For this reason, anything from a fairly smooth glass-ice glove to ragged single-ridge and double-ridge edges may form along the leading edges. Heavy, frost-like deposits may also be obtained. Fig. 10.24 shows common types of leading edge ice formations.

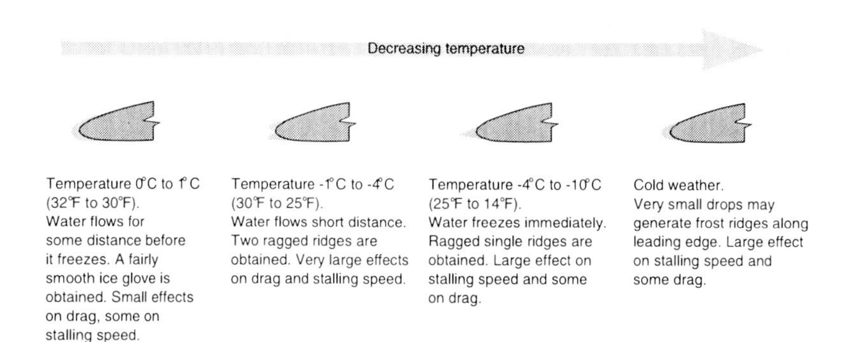

Fig. 10.24 *Leading edge ice formations*

All of the ice formations shown have considerable effects on stalling speeds and drag except the fairly smooth glass-ice glove obtained at near-freezing temperatures. However, this ice may become extremely dangerous if ice ridges are formed behind the boots when the glove is broken. It may be advisable to leave it on until the flight is out of icing conditions.

Engine Ice

Reciprocating Engines

Start-up in cold weather requires engine pre-heating. Frozen oil may overload the engine during cold starts. The result may be engine failure during take-off climb, a common cause of winter accidents.

Carburettor ice may form at temperatures between 70°F and 15° F (+20°C and –10°C) depending on the type of carburettor used. In addition to this, ice formed on the engine air intake may reduce the airflow sufficiently to choke the engine. A long and safe flight without any ice problems may quickly turn into a nightmare when the flight passes over a lake with cold haze which may rapidly give carburettor ice and engine failure.

Turbine Engines

Ice or other types of contamination on the air intake lips have the same effects on the air intake flow as wing leading edge ice has on the wing lift and drag.

Flow disturbances at the lips result in increased thickness of the internal boundary layer or air intake flow separations (fig. 10.25a). This reduces total pressure of the engine flow which again decreases thrust at a given fuel flow. It also has negative effects on engine fan blade loadings especially if the flow separations are large enough to result in compressor stalls. Erosion and foreign object damage to the air intake lips have similar effects.

The effects are largest at low speeds and high power when the angle of attack of the airflow into the engines is largest. At high speeds when the diameter of the flow into the engine is smaller than the intake diameter (fig. 10.25b), intake flow disturbances have no effect on the flow into the engine but the spill-over drag of the pod increases. Thus the total drag and the fuel consumption of the aircraft increases.

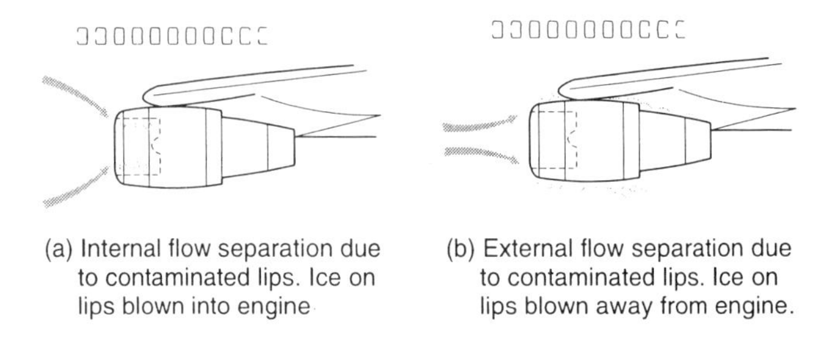

(a) Internal flow separation due to contaminated lips. Ice on lips blown into engine

(b) External flow separation due to contaminated lips. Ice on lips blown away from engine.

Fig. 10.25 *Air intake flow disturbances*

Engine air intakes must be checked for snow and ice before engine start-up. Ice sucked into the engines may damage fan blades and snow may result in flame-out, an unpleasant experience during take-off. Ice and snow ingested into engines during take-off and landing may damage or destroy engines. Here are some examples.

- The DC-9 landed in rain after a long flight at high altitude. The temperature was a few degrees above freezing. After a short turn-around time the crew taxied to the runway for the homebound flight. At take-off rotation, a sharp sound was heard and both engines began to lose power. The crew managed to stop on the runway.

  Clear ice had formed on the upper surfaces of the wings in the wing-root sections above the cold fuel tanks. When the wing was loaded during rotation the ice broke loose and was ingested by the engines. Both engines failed.

It is impossible to see the ice on the wet surface of the wing and it is difficult to feel the difference between a wet skin plate and wet ice on the wing. The best way to check for ice is to scrape the surface with a knife (without damaging the skin).

- The MD-80 had been de-iced and anti-iced before taking off on a winter day with near-freezing temperatures. Shortly after lift-off when the aircraft was in clouds, both engines failed. A passenger looking out of the window at that time saw ice blowing rearward from the wing. Fortunately, a field suitable for a crash-landing was within gliding distance when the aircraft came out of the clouds.

  Clear ice on the wings above cold fuel tanks may cause accidents or, at least, expensive engine damage for all aircraft with fuselage-mounted engines (fig. 10.26).

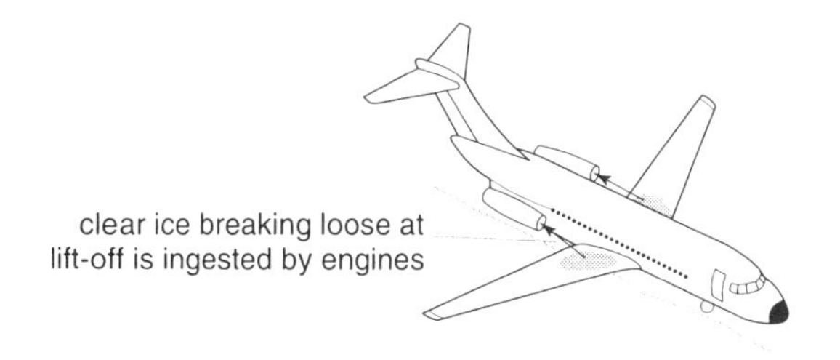

Fig. 10.26 *Dangerous clear ice above cold fuel tanks*

Ice striking fan blades may knock off pieces of the blades. This may create chain reactions when the pieces strike fan, compressor and turbine blades on their way through the engine. The damage caused by clear ice shown in fig. 10.27 is relatively small. The ice ingested at this particular take-off did not cause engine failure.

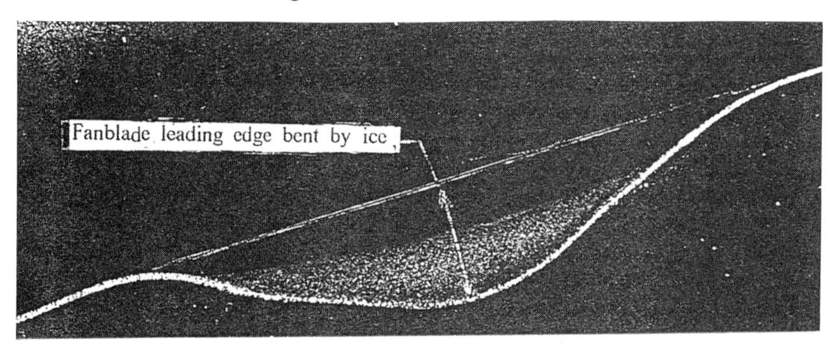

Fig. 10.27 *Clear ice damage on fan blade*

- A Boeing 727 taxied to the runway in icing conditions. When the aircraft was close to the runway the captain exclaimed, 'We forgot engine anti-ice!' The heat was switched on and the aircraft taxied into take-off position. Immediately after full power had been obtained, when the aircraft began to roll, a large bang was heard: total engine failure. Call for the tow-truck!

  Ice accretes fast on air intake lips. In a few minutes, 4–6 inches (10–15 cm) can be obtained. When heat is applied to the intake lip and full power is added, the ice is sucked into the engines (fig. 10.28). The result can be disastrous!

  Also note that ice on the engine's total-pressure probe gives false engine power readings. The indicated power is too high.

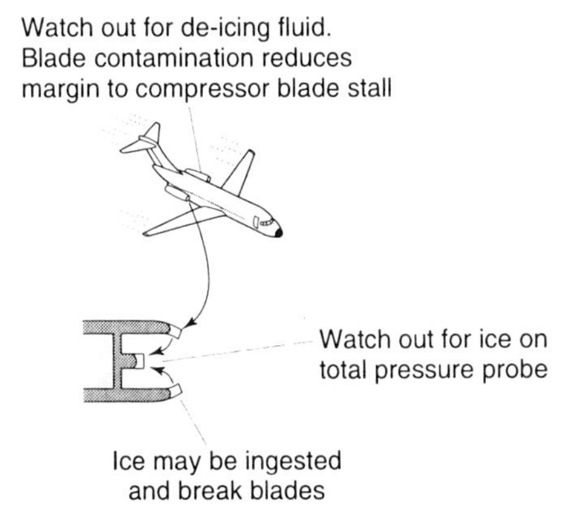

Fig. 10.28 *Ice on air intake lips may be sucked into the engines*

- A DC-9 taxied for take-off. It was snowing and the wings had been de-iced and anti-iced but there was a layer of wet snow on top of the fuselage.

  The snow remained during take-off acceleration. However, when the aircraft rotated for lift-off the snow slid rearwards and split into two streams which were ingested by the engines: double engine flame-out. Call for the tow-truck!

- The runway was full of slush and ice and quite slippery. The crew of the DC-10 used reversers until the speed was so low that slush and ice were blown forward and ingested by the engines. In this case, nothing appeared to happen and this is, of course, the case with most cases of ingestion. However, you can be sure that the compressor stalls that were induced by flow disturbances at the intake lip and the ingestion of snow and ice *may* result in engine failure or at least *may* damage engines sufficiently to cause premature engine failure or reduce aircraft performance and increase the fuel consumption.

- The DC-10 was cruising at high altitude when ice built up on an antenna on top of the fuselage. Antenna heating was switched on, the ice blew off and struck the centre engine. In a split second, the high-powered three-engine jumbo became a low-powered twin.

  Much money can be saved by those who continually stress the need to maintain smooth air intake lips and avoidance of ice or snow ingestions.

Ice on the Air Data System

Ice on the pitot tube, at the static ports, on the angle of attack vane or probe may in each case knock out the whole air data system and the automatic flight-control system of an aircraft (fig. 10.29).

- A Boeing 727 took off one winter day in icing conditions. The crew forgot to select pitot tube heating. The pitot tube froze when the aircraft had reached climb speed and sealed a high total pressure in the air data system. As a result, the indicated speed began to increase when the ambient air pressure decreased as the aircraft climbed.

  The crew increased the climb angle in order to prevent overspeed and the aircraft began to lose real indicated speed. Finally at high altitude it stalled and pitched into deep stall from which recovery was never made.

  The tailplane failed during the deep-stall descent probably due to overload by the fuselage vortex flow striking the tail section.

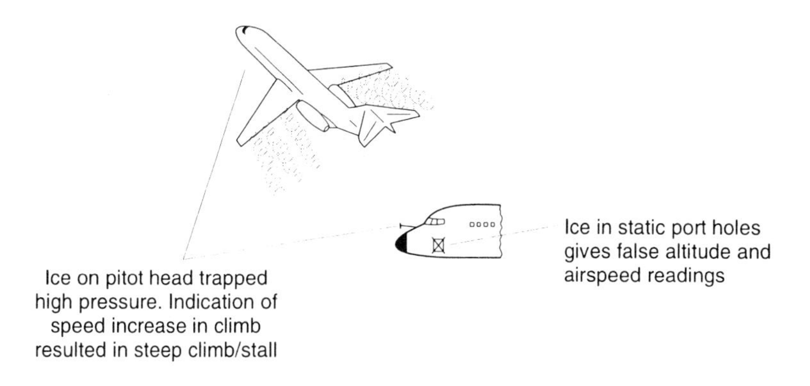

Fig. 10.29 *Air data system ice*

- A Beech Baron on approach to a small airport in instrument weather with freezing rain, struck trees a short distance from the runway just after having flown over an open field.

  The pilot survived and told the investigator that he had not observed anything wrong with the aircraft during the approach. Investigation of the instruments in the aircraft showed that the altimeters had been set at the correct air pressure and that at impact they indicated that the aircraft was flying at a safe altitude.

  The probable cause of this accident was that freezing rain, by coincidence, had slowly blocked the static pressure ports during the descent and had finally sealed too low a pressure in the air data system. Hence, the altimeters showed too high an altitude. Altimeter check by means of the alternative static pressure source (cabin pressure) would have prevented the accident.

Note that false speed and altitude indications can be obtained even in summertime when insects build nests in the pitot tubes. Also, remember to remove the tape from the static ports put there prior to washing the aircraft. The tape does not have large effects on altitude indications during climbs because the tape is lifted by the air escaping from the static tubes. However, during descents the tape is pressed against the holes of the static port by the increasing air pressure, and too high an altitude is indicated. A pennyworth of tape can destroy a jumbo in I.M.C. weather if it leads to collision with the ground.

With blocked pitot tubes and blocked static pressure ports even the most advanced electronic control systems are worthless.

De-ice and Anti-ice – Where and When

Aircraft should always be *completely* de-iced when freezing precipitations such as frost, snow, ice or freezing rain adhere to the skin surfaces, even at temperatures well above freezing when clear ice may have formed above cold fuel tanks.

If warm de-icing fluid (glycol-water mixture at 82 °C, 180 °F) is not available the aircraft must be cleaned by other means. Nobody can tell you how much contamination an aircraft will tolerate without risk of an accident especially if the aircraft runs into other problems such as engine failure or wind shear during take-off.

The clean aircraft requirement should be obvious to all pilots today. However, we still read in accident reports that crews have been offered de-icing in severe icing conditions but have rejected the offer and have done their own 'de-icing' by chipping ice off the wing leading edges of their aircraft. This is incredible! The aircraft crashed; how much money did they save?

De-icing requires very stringent rules. It is, for example, insufficient to tell the mechanic responsible for de-icing to check for clear ice above cold fuel tanks when necessary. Ten aircraft may be checked one after another without finding ice, leading to the erroneous conclusion that no further checks are necessary. However, the next aircraft may have more, and colder fuel in the tanks and ice may form. Under conditions when clear ice may form the wings must *always* be checked.

When it rains at near-freezing temperatures and aircraft are cold, long icicles may form under the wings (fig. 10.30). These may be difficult to see in darkness and special under-wing checks must be made.

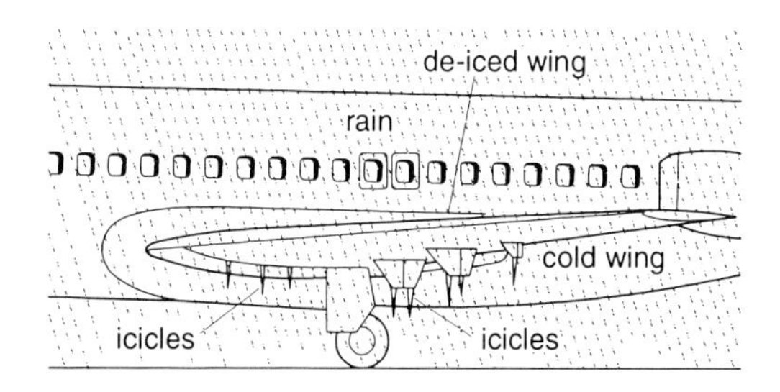

Fig. 10.30 *Check the lower surfaces of the wings*

Frost in the fuel tank area on the lower surfaces of the wings can be accepted if the temperature is well above freezing. The warm airflow melts the frost during taxying. However, the frost remains when temperatures approach freezing and should then be removed even if the risk of taking off with this frost is small. Fuel tank frost on the upper surfaces of the wings should always be removed.

Mixtures of snow and de-icing fluid form slush that freezes at high altitudes. For this reason the areas between the control surfaces and the fixed wing/tailplane surfaces should always be checked and cleaned in order to prevent control surfaces from freezing during the flight. A European airline experienced this twice one winter when the elevator froze and could not be moved (fig. 10.31). Descent to lower altitudes where the ice melted was made by means of tailplane trim.

Aircraft surfaces must be treated with anti-icing fluid in weather in which there is a risk of freezing precipitation during taxying from gate to runway.

The problem confronting us here is that the holdover time (the time for which the fluid prevents a precipitation from freezing on the aircraft's surfaces) may vary between a few minutes and several hours depending on

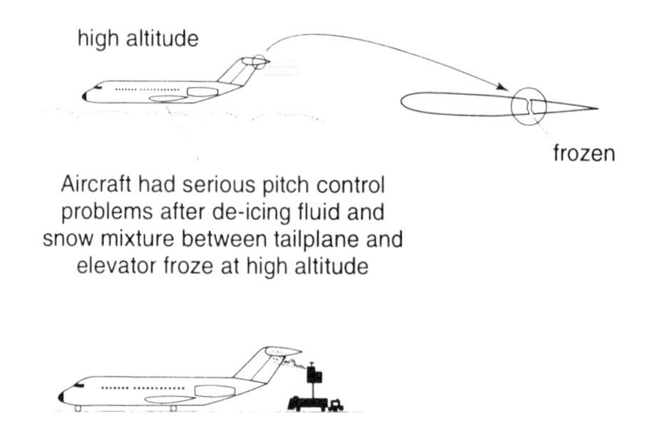

Fig. 10.31 *Elevator locked by frozen slush*

the type of anti-icing fluid used, the fluid–water mixture ratio and the weather. Snow melting on the aircraft's surfaces and freezing rain dilute the fluid and strong winds may blow fluids of low viscosity off the wings.

There are two common types of fluid in use. Type I fluid which is the de-icing fluid and often also used as anti-icing fluid. It is a hot mixture of about 80% glycol and 20% water. At this ratio, the fluid has roughly three minutes' holdover time in freezing rain if the temperature is below freezing and fifteen minutes in steady snow. This fluid is not very useful for long taxying times in severe weather.

In order to increase the holdover times, a fluid (Type II) with a thickening agent which makes it stick better to the aircraft's surfaces at low speeds has been developed. If this fluid is used in concentrated form it has twenty minutes holdover time in freezing rain, at least forty-five minutes in snow and many hours in frost.

Fig. 10.32 *De-icing/anti-icing*

These fluids are not available at all winter airports. Both types must be handled with care or they may not give the expected protection. Proper de-icing and anti-icing therefore requires well-maintained equipment suitable for the fluid used and well-trained personnel who know what they are doing.

Note that the holdover times quoted above are examples. Fluid types and mixtures vary and new fluids are under development. It is the responsibility of airline and airport management to supply aircrews and ground crews with correct holdover times.

It should be obvious that taxying times from gates to departure runways must never exceed the holdover times of the anti-icing fluids. Safe operations would, in that case, require that aircraft returned to the gate areas for renewed de-icing and anti-icing unless renewed de-icing is available very close to the runways. Returns for renewed de-icing would certainly decrease the capacity of airports in winter storms.

Since the capacity of the airport is determined by the take-off rates and not by the number of aircraft moving slowly in line for the runway, the slow motion must be avoided and replaced by reasonably high-speed taxying. This requires good co-ordination between de-icing/anti-icing, push-back, taxying and take-off clearance. At professionally managed winter airports this is what happens.

In order to reduce the risks of aircraft damage during de-icing/anti-icing the following should be remembered.

- Do not spray de-icing/anti-icing fluid into turbine engines. All fluids, especially fluids with thickening agents, contaminate fan and compressor blades and increase the risks of compressor stall. The engines should never be run at high power during de-icing/anti-icing operations and should preferably be shut down.
- Do not spray directly into the A.P.U. Fluid entering the air conditioning system is blown into the aircraft which may then have to be removed from service and cleaned.
- De-icing fluids sprayed at high pressure on flap and slat tracks remove lubrication. This increases the loads on the tracks during flap and slat movements which increases the risk of premature failure.
- Do not spray de-icing fluids on hot brakes or at high pressure into wheel wells.
- Do not spray hot de-icing fluids directly on to windows. They may craze.
- Do not spray into pitot tubes and static ports.
- Do not spray into engine outlets and reversers.

When the wings are full of ice and freezing rain is glazing everything in sight or snow is falling, it is easy to decide that 'today we have to de-ice and anti-ice'. However, sometimes it is necessary to anti-ice even if de-icing is not necessary. The following cases are typical.

- The temperature at the airport was very low. There were no signs of frost and no precipitation was expected. The crew decided that anti-icing was not necessary.

  Slats and flaps were retracted while climbing through clouds in the airport region. Slat retraction resulted in pronounced buffeting. The crew immediately lowered the nose and the buffeting stopped.

  The climb with the cold-soaked aircraft was made through an inversion layer with high humidity and ice froze on the cold, extended slats. When the slats were retracted and the airflow's forward stagnation point moved below the retracted leading edge, the ice nearly stalled the aircraft (fig. 10.33).

  Ice may build up quickly on cold-soaked aircraft climbing through humid inversion layers. The risk must be checked before departure, aircraft must be anti-iced and the slats must be heated, otherwise a serious surprise could be in store when they are retracted.

- During a cold, sunny, dry winter day there may be no need for anti-icing unless the aircraft taxying in front of you blows up dry snow which melts in the heat of the jet blast and is slammed on to your wings where it freezes (fig. 10.34).

- The risk of self-induced icing also exists when reversing or taxying in light snow (fig. 10.35).

- The Boeing 747 had been parked for a few hours on a winter's day at the large international airport. The temperature was well above

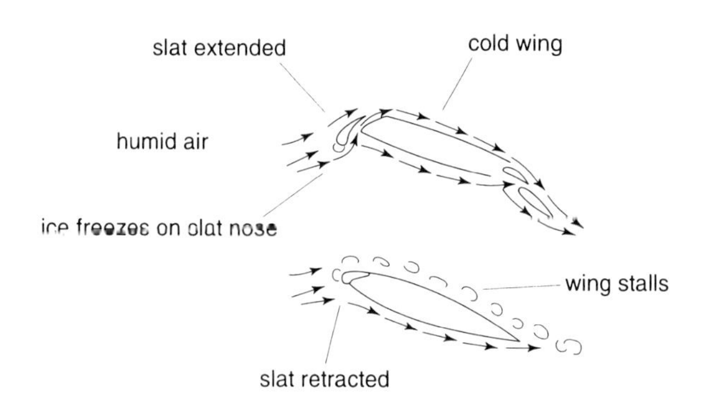

Fig. 10.33 *Ice on cold slats*

Fig. 10.34 *Light snow melts and freezes*

Fig. 10.35 *Self-induced icing in light snow*

freezing but poor weather was forecast and the temperature was falling fast. The aircraft was dry and clean during push-back in darkness and there was no precipitation. No anti-icing was done.

A minute or two after push-back snow began to fall. The snowflakes looked like small, hard snowballs which bounced up in the air when they struck the wing and then rolled off and fell on the ground.

After a while, the aircraft began to taxi at wind speed and snow began to build up on the wings. When the aircraft was in take-off position the snow layer was approximately 10 cm (4 inches) thick, but light and dry as it was, it would blow off as soon as the aircraft began to move. Or so it was believed.

The speed increased and the snow remained. At rotation, the snow was sucked away and revealed a clean surface above the wing tank but very rough ice on the leading edge flap and on the surfaces ahead of and behind the tank. The wing outside the cold tank area had been warmed during the hours the aircraft had been parked. When snow fell, heat from structures with above-freezing temperatures melted the snow which then froze when the structure cooled. The aircraft should have been anti-iced when falling temperatures and precipitation had been forecast. But with no wind shear and no engine failure, the take-off was uneventful.

- The decision to check the aircraft before take-off in precipitation and return for de-icing/anti-icing when necessary may be a difficult one but it may prevent an accident. The crew on a DC-9, which had been both de-iced and anti-iced, felt uncomfortable after prolonged taxying in freezing rain. The aircraft looked clean but the captain opened the window and felt with his hand on the fuselage side. He found that the aircraft had a fairly thick cover of clear ice.

Water Leakage Ice

Fluid leaking from toilets may seep through the fuselage skin, freeze on the outside and form a large lump of ice which may break loose and knock out an engine or damage the tailplane's leading edge.

Water leaking from pentries or rain water blown into parked aircraft through doors or poorly sealed emergency exits may flow in to wheel wells, freeze and lock control system wires.

Slush on Runways

All runway contamination decreases braking friction, even thin films of wet dust, but only high, wet grass, standing water, snow and slush also decrease acceleration to take-off speeds. Slush is worst. The wheels of an aircraft taking off from a slush-covered runway accelerate the slush to the aircraft's speed as they move along the ground. Since the force required to accelerate the slush increases with the square of the speed, the degrading effects of slush increase with increasing speed. As a result, acceleration may at first appear to be normal during take-off, but deteriorate as speed increases. During the acceleration, the landing-gear wheels may rotate backwards since the upwards-directed slush pressure on the wheels forward of the wheel axes may be larger than the rolling friction force rotating the wheels in the normal direction.

Imagine what would happen if an aircraft with contaminated wings attempted to take-off from a slush-covered runway. Since neither wing nor runway contamination effects are felt before high speed is reached, a pilot attempting to take-off with this insidious combination of detrimental factors may discover that the aircraft's acceleration decreased dramatically as take-off speed was approached. The risks of high-speed overruns becomes very large without any additional disturbance. With an engine failure, overrun is guaranteed at any speed close to V_1 since the aircraft will aquaplane on the slush during the braking attempt.

Contamination Accidents

According to the Flight Safety Foundation, between 1968 and 1991 there were thirteen jet transport accidents attributable to ice accumulation, nine in the USA and Canada, three in Europe and one in the Far East. Between 1982 and 1991, there were eight accidents, nearly one a year. If we include turboprop transport accidents and accidents in countries which do not readily report mishaps we will find that there has been more than one serious transport aircraft accident each year during the last decade or two. To this must be added all regional airline and airtaxi accidents such as the Twin Commander accident at Bromma, Sweden, caused by frost on the wings. Altogether we would find that the public transportation system has an unacceptable rate of winter contamination take-off accidents.

However, we must also consider accidents caused by in-flight icing. Turboprop transports with pneumatic de-icing systems have crashed during departure climb and cruise due to leading edge ice, and tail stall accidents continue to happen in spite of the fact that we have had six or seven Viscount accidents and a number of others.

Aircraft accelerate during take-off rotation. At lift-off the speed may be only slightly higher than the stalling speed with a contaminated wing. Nothing happens and the pilot may get the impression that the effects of contamination are small. Also, delayed and slow rotation increases the speed at lift-off and decreases the risk of contaminated wing stall.

During take-offs with all engines running, an aircraft may have no problem lifting off before reaching the end of the runway and clear obstacles along the take-off climb path. However, nearly all of the margins needed to take care of unpredictable take-off problems, such as severe turbulence at lift-off, too early a rotation, overrotation, etc., have been eliminated by the contamination. Even combinations of small detrimental factors may, therefore, precipitate an accident. Engine failure guarantees one.

The fact that hundreds of thousands of contaminated take-offs can be made before all margins have been exceeded, and the frequency of jet engine failure at take-off is extremely low, gives the impression that the risks involved are low. This permits unsafe operations to continue year after year. When accidents happen they are classified as 'unpredictable' random cases. This is self-deception.

The contamination problems should be well known but the awareness of the risks is still far too low.

A Scandinavian might tell me after reading about the risks of accidents caused by contamination that I have been 'painting the devil on the wall' because so many take-offs are made with contaminated wings without anything happening. The truth is, however, that something always happens even if the effects are not sufficient to eliminate the whole margin to stall or all of the power required for climb with all engines operating.

You can , however, be one hundred per cent sure of this: if you maintain clean and smooth surfaces you reduce the risk of accidents and you may save a considerable amount of fuel each year. It pays off!

Note: ice caused by freezing rain is extremely dangerous! *Never* fly into clouds with supercooled droplets. Get out of it fast as possible if you get caught!

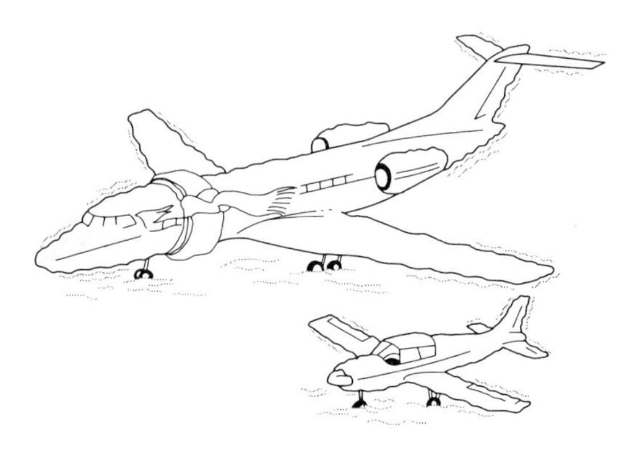

Fig. 10.36 *Large or small, keep them clean*

Actions for Improved Safety

Many suggestions have been made to reduce the risk of accident due to contamination. In case of contaminated wings, increased rotational speeds and delayed rates of rotation are used to increase the margins to stall at lift-off. However, this increases tyre wear and the risks of premature tyre failure during normal tyre life and does not solve the problem of increased aircraft drag increasing the acceleration distance to V_1 and reducing the climb rate after take-off. Neither does it eliminate the risk of stall during take-off.

The only safe way of dealing with the problems associated with the take-off of contaminated aircraft is to :

- demand de-icing near the take-off runway or co-ordination between de-icing, anti-icing, taxi clearance and take-off clearance so that take-offs can be made within safe holdover times
- inspect aircraft wings prior to take-off in summertime; wipe off insect deposits on all leading edges
- demand maintenance of smooth leading edge surfaces; erosion, dents and worn de-icing boots increase both stalling speed and fuel consumption; check for leakage from hot air ducts.

The wing contamination problem can be solved with existing methods.

A slippery runway increases the balanced field length. In order to reduce the risk of overrun, pilots sometimes reduce V_1. This solves nothing. Too little a reduction does not prevent overruns and too much of a reduction may make it impossible to lift off before the end of the runway is reached if an engine fails at V_1. Accepting take-offs from contaminated fields using dry-runway balanced field length data, for runway-limited take-offs, is the same as accepting continued take-off if an engine fails near V_1 since the alternative is an overrun crash.

The primary problem to attack in this case is the runway surface. Scored surfaces of the type shown in fig. 9.52 are recommended. Simple grooving of smooth surfaces does not give sufficient protection, especially if the groove spacing is too large.

Standing water anywhere on the runway is not acceptable. Water blown up on one side of a runway in crosswinds cannot be accepted. On single runway airports this can be prevented by erecting low fences (fine-meshed nets) along the runway at a safe distance from the runway edge. The fence slows down the wind along the ground and the water runs off the runway.

Rubber deposits at the ends of runways must be removed frequently. Aggressive runway snow and slush removal is required even if the runway surfaces are acceptable. Runway friction testing must be made at speeds up to aircraft landing speeds. This requires well-trained surface-cleaning crews and well-trained meteorologists who understand the effects of critical weather and can assess the effects of rapid weather changes on runway conditions.

All pilots must always consider the feasibility of safe take-off. A person in a general aviation aircraft facing a take-off from a water-saturated grass field must always walk the runway and check the surface condition. If in doubt, a short acceleration test can be made before take-off is attempted or take-off can be delayed.

Airline crews must decide if take-off is safe, assuming the need to abort, even with wings that are kept clean and runway surfaces that are maintained as coarse as possible under the prevailing conditions. In order to do this, they need current runway friction data and current wind and precipitation information. Data updating must be in step with changing conditions. Performance charts showing the effects of the prevailing conditions on the balanced field lengths and the need to reduce take-off weight (T.O.W.) are required to make safe take-off decisions. The following data show the corrections applied for take-off with a twin-engine jet transport registered in Sweden.

Braking Coefficient	T.O.W. reduction (kg)
0.50	0
0.45	500
0.40	1 500
0.30	4 000
0.20	10 500

Table 10.1

Contamination depth (mm)	T.O.W. reduction (kg)	
	Slush	Water
2	500	800
4	1 100	2 000
6	2 100	3 900
8	3 450	6 300
10	5 000	10 000

Table 10.2

Fig. 10.37 *Aircraft destroyed in collision with a structure beyond runway end*

Finally, it should be rather obvious that runway ends must be prepared for overruns. The aircraft in fig. 10.37 was a total loss after it collided with a strong structure in the overrun area. Compare this to the quick repair of the aircraft which knocked down the I.L.S. antenna in fig. 9.20. The worst 'solution' of the overrun problem occurred when the crew of a jet transport was jailed for manslaughter when their aircraft ran into a ditch crossing the runway extension. Airline crews had, for years, pointed out the need to cover the ditch.

Chapter 11

Risk Awareness and Failure Management

Years ago, I talked to a US Air Mail pilot flying mail between Cleveland and Boston. He told me that one day shortly after leaving Cleveland he had a solid cloud cover between the aircraft and the ground. The high vibration level in the aircraft had knocked out the clock and the compass. 'However,' said the pilot, 'I remembered that I used to smoke one cigar and down to the girdle on the second on my way to Boston. So I continued the flight, used the sun for navigation, smoked my cigars, and when I reached the girdle of the second one, I threw it over the side, descended through the clouds and there it was, Boston!' It was the time of sharp awareness and railway navigation. It was the time when the feeling in the seat of the pants and in the hands and feet integrated the pilot with the aircraft.

'I follow railways'

Fig. 11.1 *Railway navigation*

Now, automatic flight has taken over the control of civilian transports and military aircraft and as electronic equipment becomes cheaper automatic flight is spreading into general aviation. The pilot has become a system manager. This is great; it improves safety but it also creates new problems.

In the 'cyberspace' of automatic flight where all systems function

Fig. 11.2 *The system manager*

perfectly, pilots may lose contact with the real world and accept what they see on instruments as the truth because 'it comes like that out of the computer.' Pilots also have the problem of switching within seconds from long periods of low activity to highly stressed, overload situations with more to do than they can cope with when faced with sudden emergencies.

Complex systems are difficult to understand and use correctly and worse to manage in crises. Recent checks have revealed the following problems.

- A pilot admitted that he did not know where he was but said it did not matter since he had a G.P.S. (Global Positioning System) set.
- A pilot brought in his autopilot for repair. He changed altitude with the autopilot engaged on altitude hold and was worried when the autopilot automatically brought the aircraft back to the selected altitude.
- Eighty per cent of Boeing 737-300 pilots did not know that the autothrottle had to be disconnected during aborted take-offs at or above 40 knots in order to prevent automatic full power after manual pull-back.

- Seventy-five per cent did not know when the go-around mode becomes available during approach.
- Between 55% and 85% of the pilots were unaware of the functioning of various important automatic approach modes.

There is no such thing as completely safe automatic flights. Safe flying requires awareness of the 'real world situation'. Automatic systems are flying aids. The pilot must develop and maintain methods for manual follow-up of the flight situation. The following examples illustrate where a better awareness of risks and of failure management could have prevented accidents.

Approach Accidents

The aircraft in fig. 11.3 crashed on an I.L.S. approach in fog. Radar observations showed that it was initially on the slope but about a third of the way down, descended below the slope. It flew into trees in a 3° descending path about 3000 ft from the runway. The accident investigation showed that a condenser in the I.L.S. receiver fly-up circuit had weakened and given the pilot a fly-down signal and a flag warning.

The main factors combining to cause the accident when the I.L.S. receiver failed were:

- a failed instrument that continued to give false information after failure which it should not have done
- insufficient warning provided by the flag
- lacking awareness of failure risks the pilot did not scan his instruments to monitor the approach; he had two receivers.

An aircraft on I.L.S. approach during darkness in falling snow crashed in a 30° bank to the left of the approach path (fig. 11.4). Radar observations showed that the aircraft was on the localiser and descending towards the glide slope when roughly 2000 ft from the runway it suddenly turned left and flew into the snow-covered ground.

The investigation showed that a counterweight in the horizon gyro had loosened and hit a screw during turbulence. This made the gyro tilt and indicate a right turn. The pilot 'levelled' the wings and turned away from the approach path. There was no warning for this failure.

The main cause of the accident was the pilot's lack of failure risk awareness and hence lack of instrument scanning. Information was available to tell him that the aircraft was still on the right track.

A heavy transport was on an automatic I.L.S. approach to a major airport when instruments showed that the aircraft was drifting to the right

Fig. 11.3 *Crash caused by failed I.L.S. receiver*

Fig. 11.4 *Approach crash in darkness and falling snow*

of the track. The crew classified this as instrument failure and continued the automatic approach. The aircraft nearly collided with the terminal building as a pull-up and go-around was made. The crew attempted a second automatic approach with the same result (unbelievable) before a third, manual landing was made.

Lack of situation awareness and blind faith in the computers have caused many approach accidents. In three fairly recent cases, crews have failed to monitor approaches and let large, well-equipped modern transports fly into the ground due to faulty system management. What 'comes out of the computer' is only correct if the input to the computer is correct and the computer has not failed.

Danger Warnings

The crew of the aircraft in fig. 9.18 had a flashing sign on the glareshield announcing 'Not ready for take-off' because the hydraulic control system had not been switched on. They did not observe the warning and did not react. The problem with the warning was that it also warned when the nose wheel steering was set at ±12° for taxying (at take-off it was switched to ±2°) and therefore began flashing as soon as the engines were started and kept on doing so during taxying. The continuous flashing made the crew 'sign-blind'.

However, full control movements must be made to check free control movements before every take-off. The crew did not do this because 'they had done that on the previous take-off'.

Repeated false warning has made crews disregard real warnings in situations where immediate action would have prevented disaster.

A heavy transport was on a night approach over 'black hole terrain' when the first officer misread the altitude setting over the outer marker (900 ft too low). The captain set the wrong altitude on the altitude pre-select and took a 45° shortcut to the O. M. During the descent the G.P.W.S. (ground proximity warning system) started sounding 'woop, woop, pull up, pull up'. 'Bueno, bueno,' said the captain and let the aircraft descend into the ground. A total lack of situation awareness makes a crew classify a genuine warning for collision with the ground during a night approach as a false warning. In cases like this, there is only one safe action: pull up and check the system later.

A modern 'computerised' transport was on a departure climb when the crew had an in-flight engine reverser warning. High power thrust reversal in flight is extremely dangerous. It destroys wing lift and yaws the aircraft into a roll. The crew discussed the problem for some time wondering if the warning could be false and caused by humidity. The engine reversed. The aircraft yawed and rolled and broke up in the air.

These are warnings that must be heeded immediately because a delay may cause a disaster.

Who Is Flying the Aircraft?

'The captain was flying the leg but nobody was flying the airplane' someone wrote in N.A.S.A.'s *Feedback* in connection with an incident where the crew did not follow up where the aircraft was going. In spite of automatic flight, someone must always pay full attention to the flight instruments. If not, the following may happen.

- A single-engine four-seat general aviation aircraft was on take-off climb on autopilot through clouds when a failure made the aircraft bank to the right and descend. The pilot did not observe what happened and failed to pull up.
- A modern heavy jet was on automatic climb-out through a snowstorm when the left engine spooled down and the aircraft rolled to the left. The crew did not observe the roll and loss of altitude before it was too late to pull up.
- The jet transport was on a night approach over water when 'gear down' was selected. When the nose-gear did not indicate 'down and locked' the whole crew began to check what was wrong. The transport flew into the ocean while they were too busy to fly.
- During a night approach to Miami the crew of the jumbo did not get a nose-gear 'down and locked' indication, discontinued the approach, flew to the Everglades area, set the aircraft in a circling altitude hold and began searching for the fault. By accident the 'altitude hold' was disconnected and the aircraft began a slow descent to the dark swamp jungle. The first officer discovered the problem eight seconds before the crash. However, it takes up to eighteen seconds for pilots to begin scanning, catch up with instrument movements and take action.

In automatic flight, someone must always be able to answer the questions 'Where are we now and what is the aircraft doing?'

Dangerous Failure Survivals

Alert, well-trained crews 'in the loop' with the aircraft and knowing their aerodynamics can control and land even severely damaged aircraft.

Fig. 11.5 *Is survival possible after wing failure?*

Examples:

- Aircraft with pitch control problems have been landed by combined use of engine thrust variation, flaps and spoiler extensions. It is possible to analyse the problem and describe optimum procedure.
- Aircraft with limited roll control can be controlled in roll by yawing the aircraft with the rudder.
- Differential thrust has been used for yaw control in cases of fin or rudder failures. A Convair 880 was landed with differential thrust after losing two-thirds of the fin.
- Since wing lift increases with decreasing sweep angle it is possible to prevent uncontrollable roll of swept-wing transports which have lost outboard wing sections by yawing the damaged wing forward. Successful landings have been made in several cases.
- A heavy jumbo freighter was climbing after take-off from Schiphol Airport (Amsterdam, Holland) when no. 3 engine broke loose, pitched forward, was swept outboard and knocked off no. 4 engine. The engines took with them the wing leading edge forward of the front wing spar from inboard of the no. 3 engine to outboard of no. 4 (fig. 11.6).

  The instruments do not show what really happens and it is impossible to check the damage from the flight deck, especially in an emergency situation. However, crews should be trained to suspect that instantaneous loss of two engines on the same side probably means that the engines have fallen off and when they do so they rip off part of the wing leading edge.

  The flight continued as shown in fig. 11.7. The crew began a return to the airport, requested runway 27 but came too far south and was sent on a 360° turn to position for the runway. Important events and communication during the flight are shown in fig. 11.7.

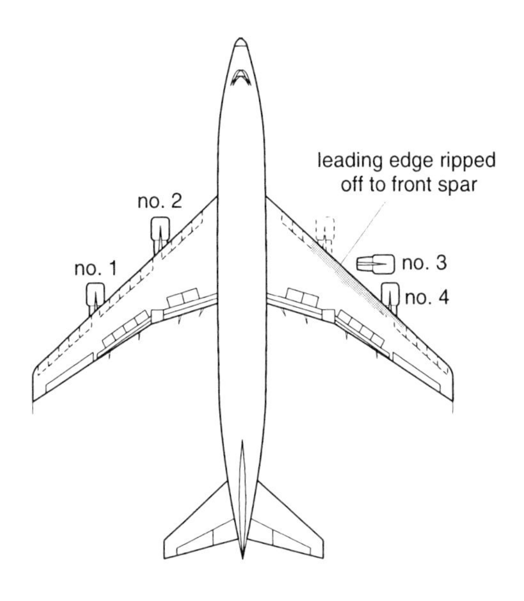

Fig. 11.6 *Loss of engines and wing leading edge*

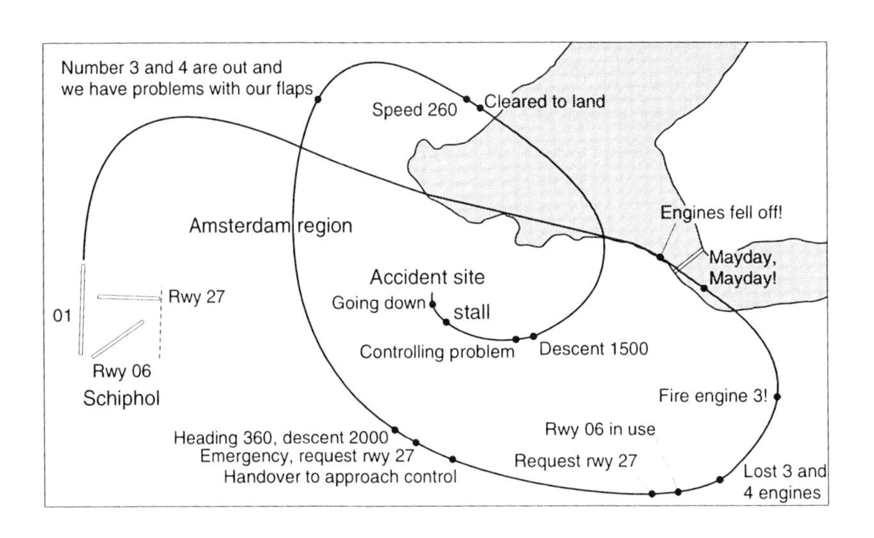

Fig. 11.7 *Flight path with lost engines*

During the descent and deceleration for the landing, the aircraft rolled to the right, turned away from the approach track and crashed into a high apartment building in a 90° bank. The reason for the uncontrolled roll is obvious. With a damaged leading edge the right wing would stall at a speed higher than the normal approach speed. Hence, the handbook could not be used to select approach and landing speeds. Also, with the high drag of the damaged right wing and high thrust on the left wing engines plus damaged hydraulic systems (reduced rudder control, floating right wing aileron, etc.) yaw control could not be maintained at low speed. Even the slightest analysis would have revealed the problems. Engines had fallen off before. The risks were known and crews could have been trained to deal with the problem. The aircraft flew for more than four minutes after engine loss before it crashed, so it was flyable provided speed had been maintained.

There are several cases on the record where crews have been able to land aircraft after losing wing-mounted engines and with damaged wing leading edges. The successful landings were not always the result of outstanding crew training. Voice recorder readouts show that the captains asked for handbook approach speeds but could not slow down the aircraft due to control problems and had to make high-speed landings. A bit of luck saved the aircraft.

Risk awareness, flight situation awareness, and training for complex failures in a complex world prevent accidents. Accident data for analysis and training are available!

CHAPTER 12
HELICOPTER PROBLEMS

Helicopter Safety

The vast majority of helicopter accidents are caused by lack of understanding of helicopter aerodynamics, stability and control, design limitations and operational risks.

Better pilot/operator training and increased risk awareness could considerably improve safety and operational efficiency. It is typical that well-trained pilots flying heavy helicopters in offshore operations have an accident rate of between one and two per 100 000 flight hours, while less well-trained and experienced pilots flying light, piston-powered helicopters have an accident rate ten to twenty times higher. Typical helicopter safety problems are:

- loss of main rotor lift and flapping control
- loss of tail rotor force/control
- hard landings, nose-over and roll-over
- collisions with ground, wires, trees and other objects
- loss of external references
- flights into dangerous weather/winds
- fuel exhaustion
- engine failure
- rotor failure
- airframe failure.

All problems, from loss of main rotor lift to fuel exhaustion, are pilot related. Even engine failure, rotor failure and various types of airframe failure can be traced to lack of knowledge of design limitations.

Main Rotor Lift and Thrust Control

The rotors rotate at constant speed. Lift is changed by changing the blade angles with a linkage system from the collective control via a fixed and a rotating swashplate to the blades (fig. 12.1).

Thrust in the desired flight direction is obtained by tilting the rotor with the cyclic control. This tilts the swashplates which changes the blade pitch angles during rotation. The angles of the forward-moving blades are decreased and the blades therefore sink as they move forward. The angles of the rearward-moving blades are increased and the blades climb as they move rearwards (fig. 12.1).

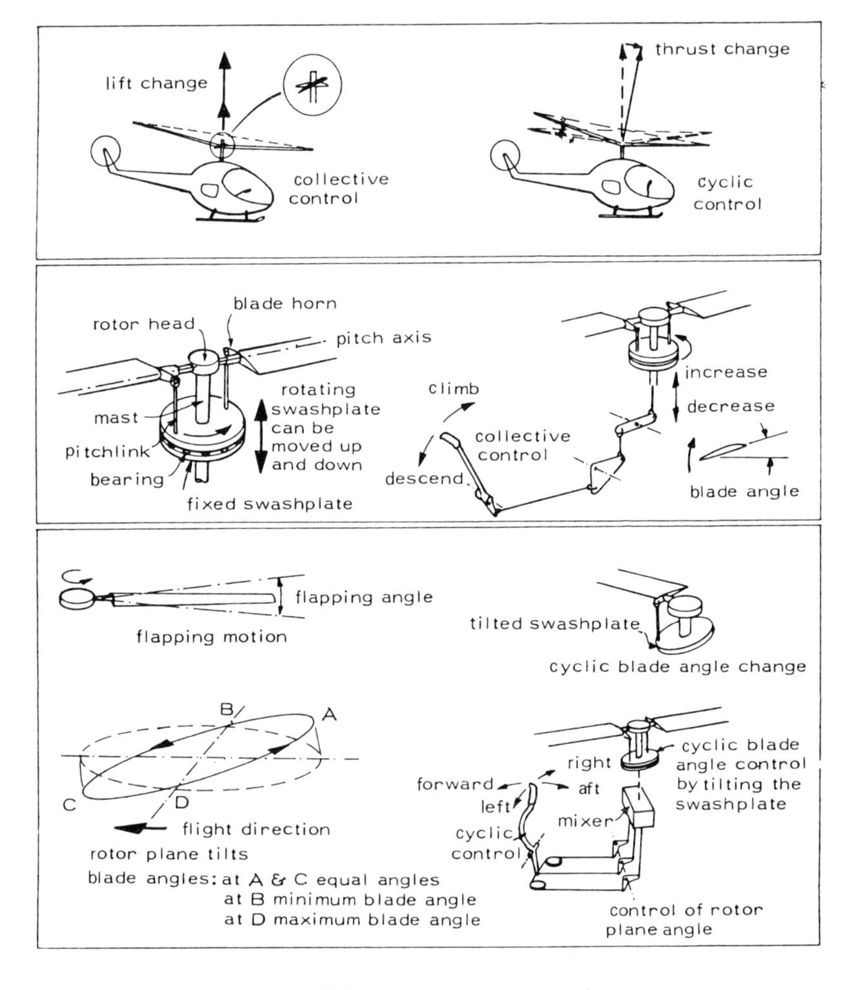

Fig. 12.1 *Lift and thrust control*

Loss of Rotor r.p.m. – Autorotation

For helicopters with manual control of r.p.m. loss of rotor speed is a serious safety problem. Inattention to rotor r.p.m. during flight may result in dramatic loss of lift. Since lift is proportional to dynamic pressure, it decreases twice as fast as r.p.m. (fig. 12.2a). If the r.p.m. loss is not immediately checked the rotor centre section will stall as the helicopter begins to fall. The stall may spread along the blades and make it impossible to control blade flapping or to increase rotor r.p.m. With lift reduced below a certain level, the helicopter automatically goes into a dive from which recovery is impossible.

Engine failures always result in loss of r.p.m. even if the collective control is lowered as soon as the engine stops. The r.p.m. lost depends on the pilot's reaction time. Rotors with low inertia lose speed quickly but spin up quickly if the r.p.m. loss is checked in time. High inertia rotors lose r.p.m.

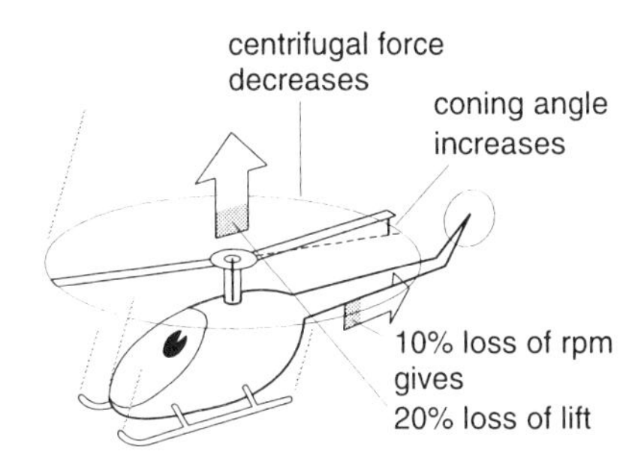

(a) Lift is lost twice as fast as r.p.m.

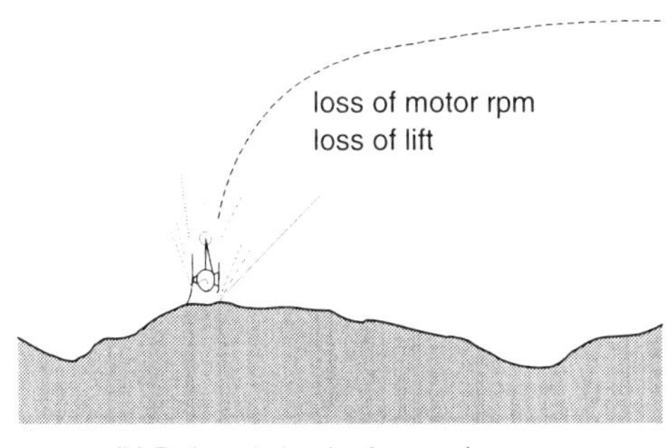

(b) Delayed check of r.p.m. loss may result in uncontrollable dive

Fig. 12.2 *Risks of r.p.m. loss*

more slowly and give pilots longer reaction times. However, heavy rotors regain rotational speeds more slowly than light ones.

When a helicopter with a failed engine descends in autorotation, air flowing up through the rotor centre section powers the rotor and drives it round. Near the rotor hub, however, the blades are stalled due to the high local angle of attack caused by low velocity of the profile and a high speed upward air flow. The upward flow through the rotor's power section is deflected downward again by the outboard ends of the blades.

A sketch of the flow through a rotor during autorotation is shown in fig. 12.3a. The relative sizes of the rotor's power area and propeller area in autorotation at forward speed is shown in fig. 12.3b.

Due to the r.p.m. losses and the altitude needed to 'fall' and regain r.p.m. helicopters have an altitude–speed limit from which safe autorotation landing cannot be made following engine failure. A typical 'dead man's curve' is shown in fig. 12.4. At low altitudes the helicopter may be landed before the rotor r.p.m. decays too much. At intermediate altitudes the r.p.m. losses result in extremely hard landings and at high altitudes there is sufficient time to regain r.p.m. for safe autorotation landings. At high forward speeds the kinetic energy of the helicopter may be used to regain lost r.p.m. (by reducing the blade angles and tilting the rotor rearward). For this reason, the low, unsafe altitude increases and the high altitude limit decreases with increasing speed until the two meet.

Note that the 'dead man's curves' published in flight manuals have been verified in flight tests by very proficient test pilots knowing what is going to happen. Flight tests have shown that ordinary pilots caught by unexpected engine failure need roughly 50% more altitude and 50% more speed to regain safe r.p.m. for autorotation landing.

The following may be said about autorotation.

- Regular autorotation training pays off.

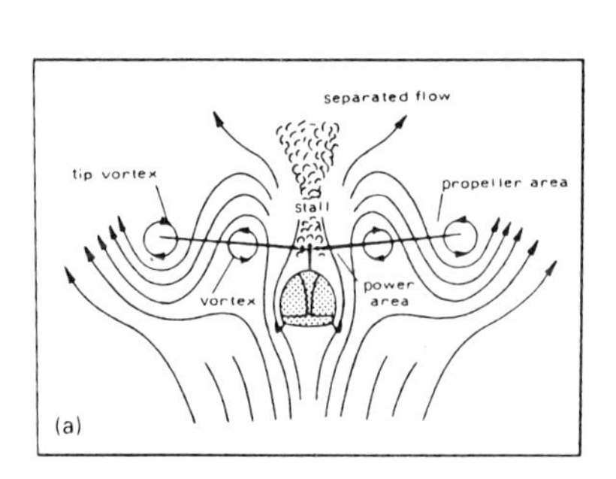

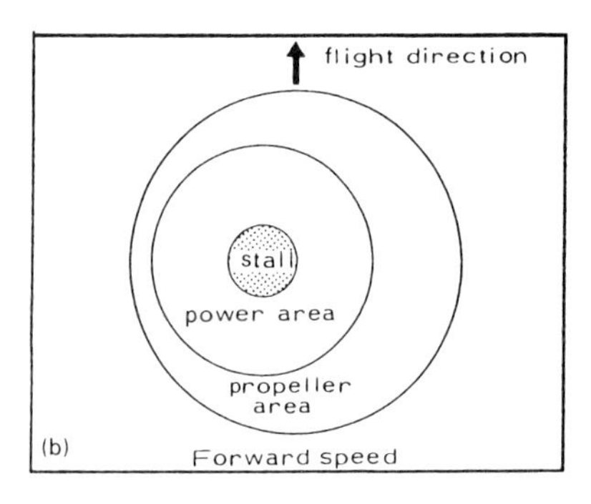

Fig. 12.3 *Autorotation*

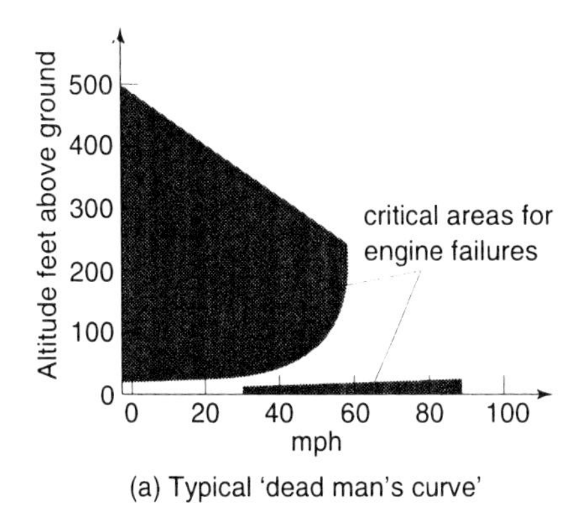

(a) Typical 'dead man's curve'

(b) Testing autorotation limits

Fig. 12.4 *The 'dead man's curve'*

- Never train for autorotation landings in strong winds, turbulence or reduced visibility.
- Land on airports or other large, open, flat grounds; avoid rough ground and areas with wind disturbances from buildings, forests, etc.
- Always land into the wind; in crosswind landings at moderate forward speed the helicopter may weathercock as rotor r.p.m. decays (reduced tail rotor trim), slide sideways and roll over.
- Pilots flying over flat terrain or water have 80% to 90% successful autorotations; pilots flying low over rough terrain have roughly 50% successful autorotations; the area available for landing increases more than twice as fast as the flight altitude depending on time required to establish stabilised autorotation glide.
- Maintain the gliding speed shown in the flight manual; attempts to

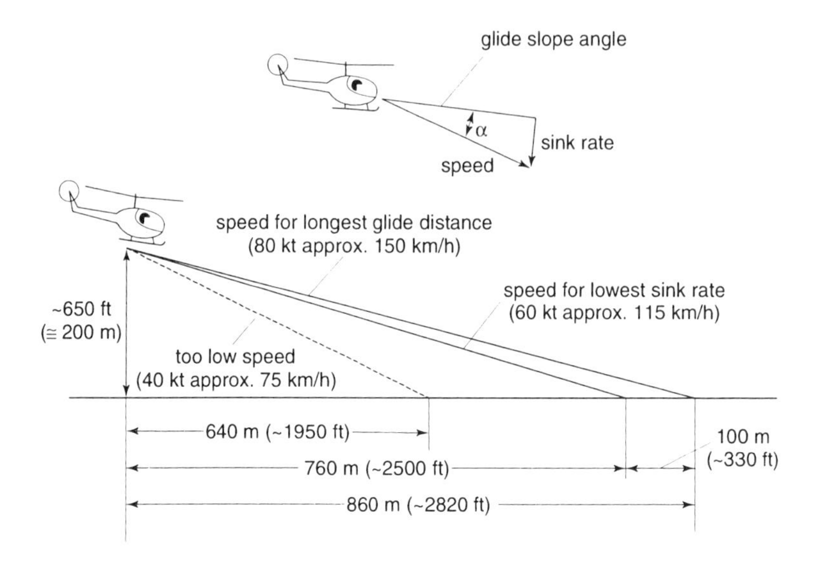

Fig. 12.5 *Autorotation glides*

stretch the glide in order to reach a landing spot results in loss of speed and rotor r.p.m. resulting in reduced gliding distances and hard landings (fig. 12.5).

- Whenever possible make take-offs and climbs over areas where autorotation landings can be made.

Recirculation and Vortex Rings

Theoretically, helicopter rotors create lift by accelerating air from infinity above the rotor to infinity below (fig. 12.6a). However, in real flow the slipstream blown downward by the rotor is slowed by the surrounding air, spreads outward and may under certain conditions turn upwards towards the low pressure region above the rotor and recirculate through the rotor, as if the rotor were enclosed in a cylinder with bottom and top surfaces (fig. 12.6b).

Rotor blades creating lift shed tip vortices. These vortices flow downwards with the slipstream as shown in fig. 12.6c. If the helicopter descends vertically or at a fairly steep angle at a moderate rate of sink with fairly large blade pitch angles it 'catches up' with the descending tip vortices. The vortices combine into a vortex ring accentuated by upflow of the surrounding air. The rotor no longer accelerates air from 'infinity' above

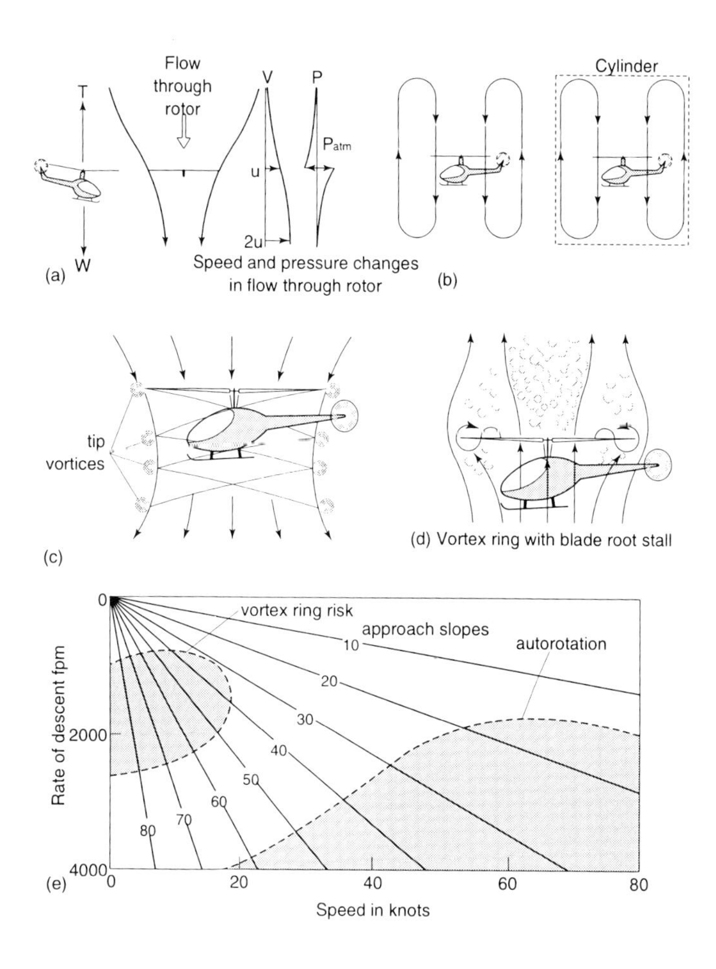

Fig. 12.6 *Flow through rotor, vortex ring and autorotation*

to 'infinity' below. Engine power is expended in keeping the vortex ring rotating (fig. 12.6d). The helicopter loses lift and enters an accelerating descent. As a result flow from below breaks through the rotor centre and stalls the blades. Any attempt to stop the descent by raising the collective (increasing the blade angles) increases the stall which spreads outward on the blades reducing lift and increasing rate of sink. The power area decreases and rapid r.p.m. decay may be expected. Recovery from a vortex ring state can only be made by lowering the collective to reduce the blade angles and pushing the cyclic forward to dive away from the vortex ring. Recovery requires immediate and firm action to prevent blade stall from spreading. Blade stall may delay or prevent flapping control. Typical speeds and sink rates for vortex ring development and autorotation are shown in fig. 12.6e.

Near the ground, recirculation can be obtained in hovering flight. When a helicopter hovers above a smooth surface the slipstream flows radially outwards three rotor diameters before it turns upwards and begins to re-circulate (fig. 12.7a). However, above high grass, close to bushes or near a wall, recirculation may begin immediately (fig. 12.7b). Asymmetrical recirculation changes the angle of attack distribution in the rotor disc (fig. 12.7c). This tilts the rotor without any control movements by the pilot and the helicopter deviates from the intended flight path. Quick asymmetrical recirculation obtained when flying along the leeward side of a forest in a strong crosswind may give serious control problems.

An impression of the size of helicopter vortices is given in fig. 12.7d, which shows a helicopter making a pull-up during a spray run close to the ground.

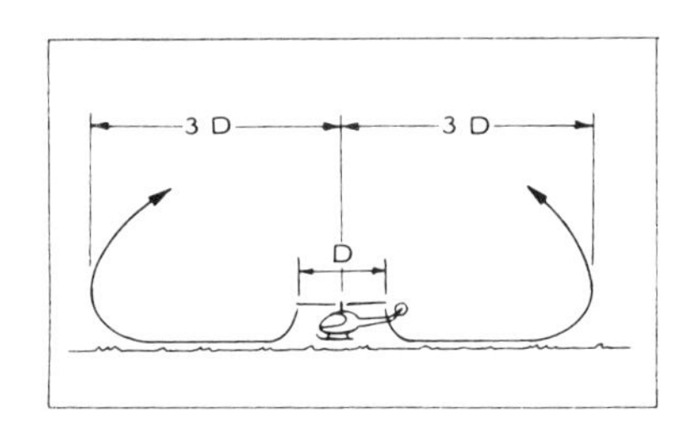

(a) Slipstream flow above smooth surface

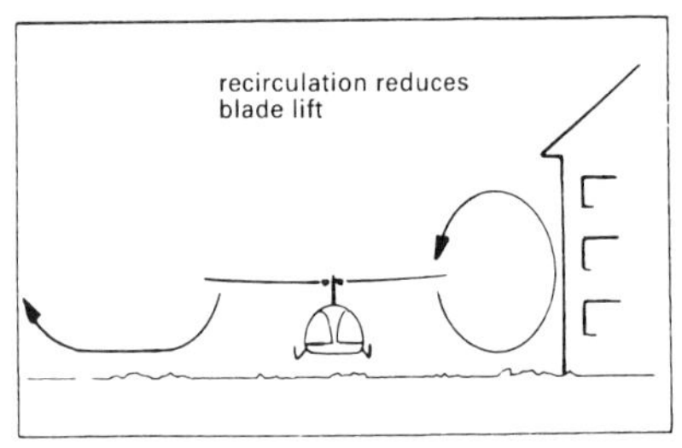

(b) Recirculation near a wall

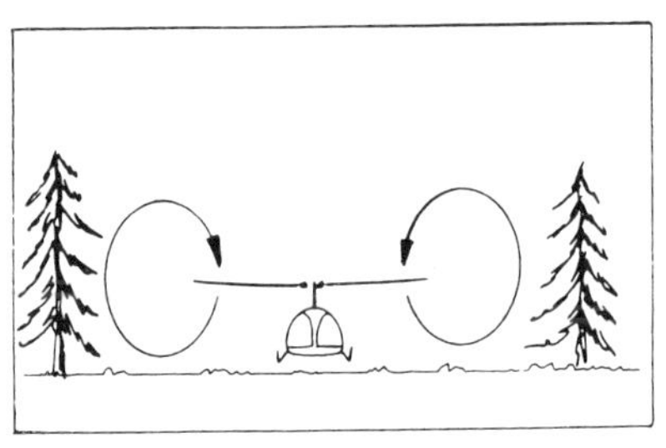

(c) Recirculation between trees

(d) Vortices during pull-up near ground

Fig. 12.7 *Slipstream flow and recirculation close to ground*

Loss of Flapping Control

The maximum speed of a helicopter is limited by retreating-blade stall and shock wave-induced stall on the advancing blade when the blade tips reach transonic speeds (fig. 12.8a and fig. 12.9a). The transonic shock stall can be delayed by sweeping the blade tips rearwards or by reducing the blade chord's relative thickness by means of a paddle shape (fig. 12.9b). The retreating blade stall region in the rotor disc is shown in fig. 12.8b.

When the retreating blade stalls it may, instead of climbing, flap down sufficiently to strike and cut off the tailboom. The risk increases with increasing weight, in manoeuvres with increasing load factors and in turbulence.

Transonic stall of the advancing blade tips shake the blades. If the stall spreads too far inboard, the blades may descend more then required for level flight, the rotor tilting forward causing the helicopter to enter a dive.

The risk of uncontrolled high-speed blade flapping increases in turbulence. Severe, sudden updraughts may stall a rotor blade on a helicopter in high-speed cruising flight. High-speed flight through heavy turbulence should be avoided. The risk of uncontrollable flapping increase dramatically with loss of rotor r.p.m.

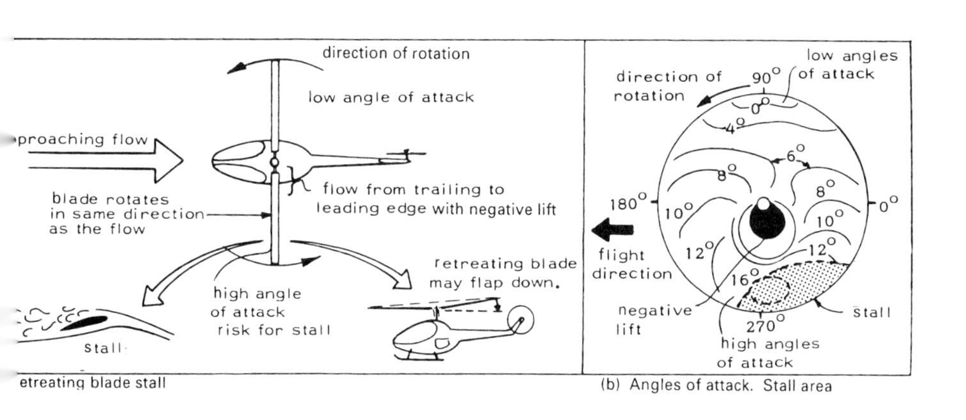

Fig. 12.8 *High-speed blade stalls*

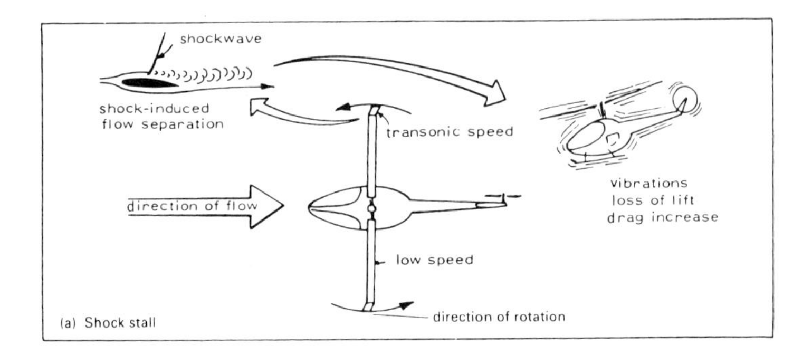

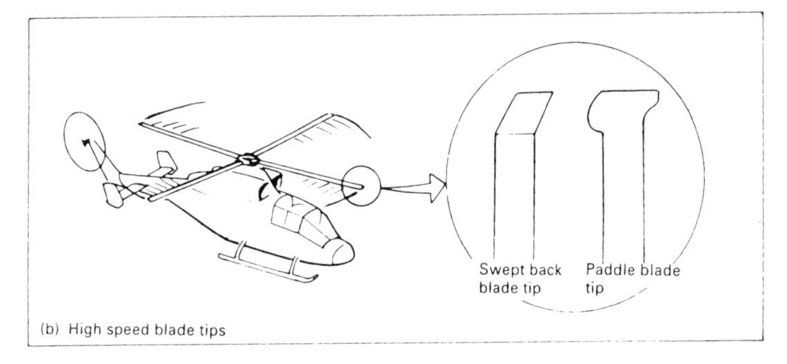

Fig. 12.9 *Transonic shock stall on advancing blades*

Mast Bumping

Helicopters with see-saw rotors may run into mast bumping when flapping control is lost and the flapping limits are exceeded. The static stops in the rotor head may strike the mast with sufficient force to make it fail. The resulting crashes are always fatal.

Mast bumping also occurs in low *g* push-overs during terrain following flights. Normally the rolling moment imposed on the fuselage by the tail rotor is balanced by an opposing main rotor moment obtained by tilting the main rotor in the opposite direction to the tail rotor thrust. However, when the main rotor thrust is reduced to zero in a bunt, the tail rotor rolls the fuselage. If the pilot now counters the roll with an opposing cyclic control movement, without increasing rotor lift, the rotorplane tilts in the opposite direction of the fuselage roll, the flapping limits are exceeded and the static stops strike the mast sufficiently hard to break it (fig. 12.10).

For this reason, push-overs to low load factors should never be attempted in helicopters with see-saw rotors, especially light ones. The risk of fatal mast bumping is too great. Large negative flight path changes can only be made safely by first reducing the flight speed to one permitting a downward flight path change without reducing the load factor very much

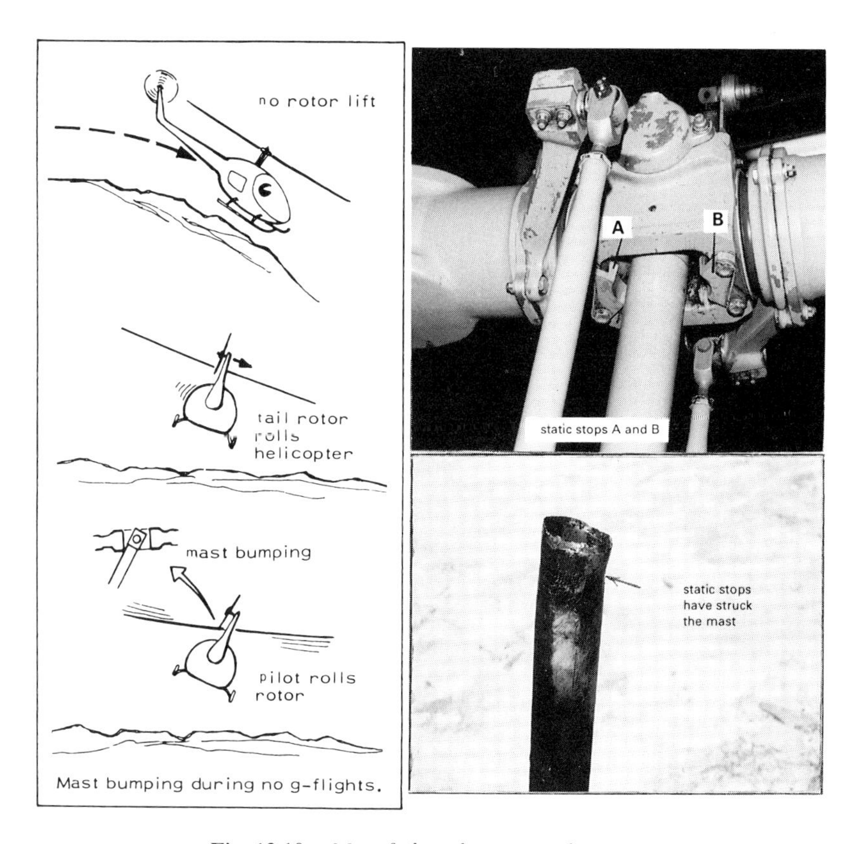

Fig. 12.10 *Mast failure due to mast bumping*

below 1 g. If speed is maintained, a very long path close to 1 g is required (fig. 12.11). Helicopters with more than two blades have no static stops and no risk of mast bumping. However, if too violent a push-over is made, the blades may contact the fuselage.

Rotor Wear and Contamination

All rotor blade wear and contamination increase drag and reduce maximum lift. From a fuel consumption and operational economy point of view, therefore, it pays to maintain the smoothest possible blade surfaces. The cost of wiping blades clean is considerably lower than the cost of operating with contaminated blades.

Flights in icing conditions often result in severe vibrations when ice is unevenly shedded from the blades. Ice growth on the blade is indicated by a need for an increase in power for level flight at constant speed. This may

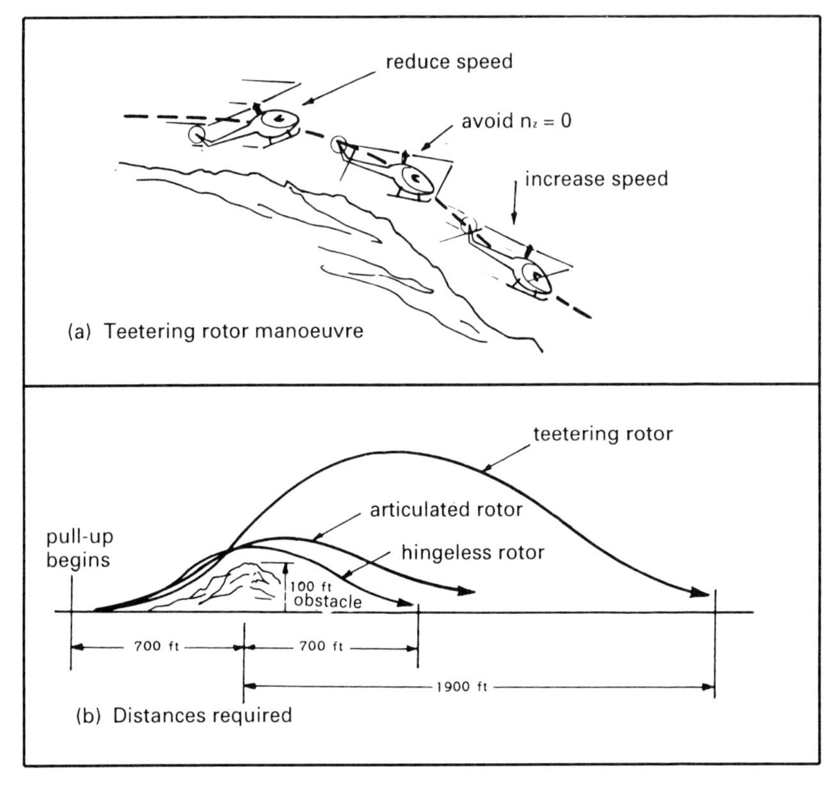

Fig. 12.11 *Push-over manoeuvres with various rotor types*

happen *before* any ice is shown on the helicopter's windshield. The increased rotor drag may have a large effect on range.

As a result of the reduction of the blade stall angle caused by ice accumulation, retreating blade stall, loss of flapping control and blade strikes on the tailboom may occur at normal cruising speeds. During flights in icing conditions engines may flame-out due to ice or snow ingestion. With a contaminated rotor, autorotation landing is not necessarily possible. Blade contamination increases the stall in the rotor disc centre. This reduces the power section driving the rotor in autorotation and a higher sink rate is required to maintain 100% rotor r.p.m.

The increased sink rate increases the stalled area in the rotor disc centre which further increases the sink rate. The result may be a very hard landing or a total loss of control.

However, even with no engine failure, ice formation may quickly make continued flight impossible and gentle, precautionary landings difficult. This is shown by the following cases.

- The helicopter was hovering in a headwind immediately below a cloud base in preparation for a landing on a mountain plateau. During the hover, the rotor picked up some ice. After a few minutes the pilot headed for the plateau. When the helicopter entered the downdraught at the plateau's edge, the contaminated rotor could not stop the descent and the helicopter struck the plateau's edge.
- During a late-autumn flight along a narrow, winding river the helicopter suddenly flew into supercooled haze above a frozen lake. The pilot knew that continued flight would be impossible, aimed for a straight piece of road, landed at high forward speed and slid to a stop (he fortunately missed the ditches and the telephone poles). The rotorblades were covered in ice.
- The pilot stood parked near the shore of a lake. High hills surrounded the lake. It was late autumn. The sun was shining. Light frosty haze lay over the open lake surface.

  The pilot decided to make a turn over the lake to gain speed and altitude for his flight to a destination beyond the hills. After the turn, the helicopter refused to climb and descended to a very hard landing at the departure field. It iced down in a few seconds.

Ice is not the only contamination which can create problems for helicopter pilots. Dead insects smeared along the leading edges of the blades can reduce the lifting capacity of a heavy helicopter by several hundred pounds and make climbs over obstacles impossible. The pilot of the helicopter in fig. 12.12 left his helicopter with the rotors running at idle power for fifteen minutes in a mosquito-infested clearing in a forest. When he lifted off, accelerated along a road and tried to climb over powerlines, which he had done the day before without any problem (but with rotors stopped while parked), the rotor lost r.p.m. The pilot made a quick 180° turn and the helicopter fell down in a ditch. The accident investigation

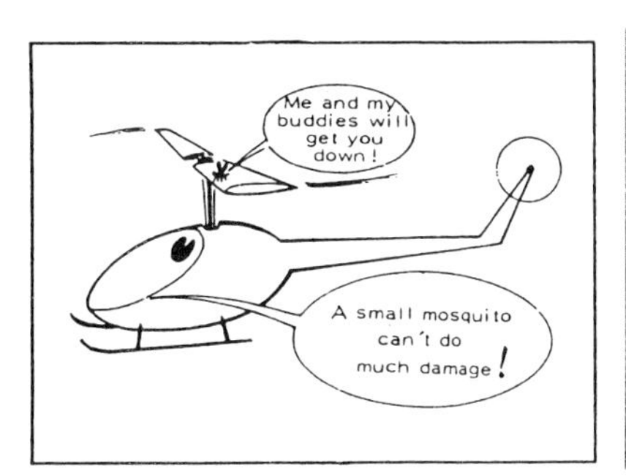

Fig. 12.12 *Insect-contaminated rotors gain drag and lose lift*

showed that the rotor was heavily contaminated with dead insects. If you believe that a mosquito causes no harm you are wrong (fig. 12.12).

Salt contamination on rotor blades and in gas turbine compressors may have large effects on helicopter performance. This is illustrated by the following incident.

The pilot was flying (hovering) in a 70-knot wind above a sinking fishing boat in the North Sea hoisting the boat crew to safety. After the rescue, when the pilot tried to fly towards the shore against the wind, he found that the helicopter would not move at maximum power. 'Fortunately,' the pilot said, 'a rain cloud drifted by and washed us clean. Thereafter we had no problem flying to the shore.'

Helicopters operating in salty atmospheres must have rotors and engines cleaned once a day. Flight tests have shown that blade contamination increases blade vibrations and that this may induce large vibration loads in the pitch links. The result may be premature control system failure. Helicopters must be kept clean. Contamination creates rotor system problems. Filth in the fuselage promotes corrosion and fatigue failures.

Fig. 12.13 *Keep helicopters clean*

Loss of Tail Rotor Force and Control

At high speeds, the tail rotor can be off-loaded and the main rotor moment balanced by the helicopter's fin. At low speeds, however, the main rotor moment yaws the fin beyond the stall limit and directional trim and control can only be maintained by means of the tail rotor.

Under certain conditions, such as combinations of maximum main rotor torque, increased density altitude and low speed, the tail rotor operates close to its maximum thrust limit and may 'stall', especially if it is worn or contaminated.

When the maximum thrust of the tail rotor is exceeded the helicopter begins to spin in the opposite direction to the main rotor. As a result, a vortex ring is formed on the tail rotor, the tail rotor thrust is reduced and the spinning motion increases (fig. 12.14). Now it is impossible to stop the spin by kicking opposite pedal. The increased tail rotor blade angles increase the blade-root stall which now spreads from the hub towards the blade tips. Recovery can only be made by reducing the main rotor torque (autorotation) combined with an increase in forward speed. This requires altitude; recovery at low heights above the ground may be impossible.

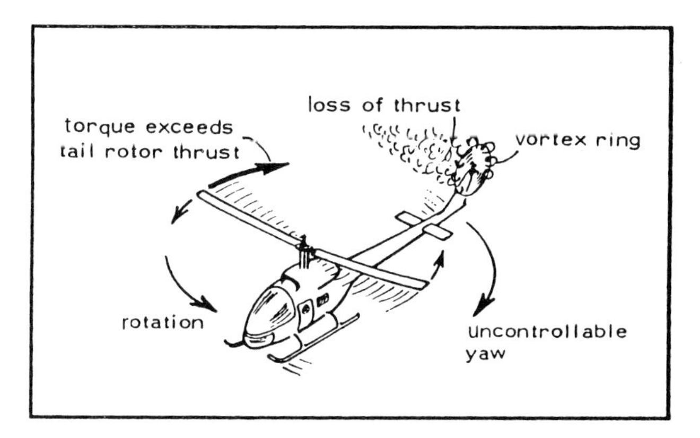

Fig. 12.14 *Tail rotor stall*

Power line inspection in a tailwind with a two-seat light helicopter, with a pilot and observer on board, is a high-risk operation with respect to tail rotor stall (fig. 12.15). In a tailwind the air speed may be low and at low speed the rotor drag is high (large blade angles and high induced drag). At medium altitudes the excess thrust available for manoeuvres may be low and maximum available power may be exceeded. Turbulence and cross-wind disturbances may start a spinning motion that cannot be stopped, especially if the helicopter loses r.p.m. when maximum available power is exceeded. Tail rotor thrust is lost twice as fast as r.p.m. Avoid power line inspections in tailwinds!

In hover and at low speeds the tail rotor operates in a very complex flow environment. It is affected by the helicopter's motion (forward, sideways or rearward), by the downsweep and recirculation of the main rotor slip-stream and by the yawing motion of the helicopter.

The problems associated with hover in various wind directions are illustrated in fig. 12.16. The helicopter is directionally unstable in a tailwind. The instability may be impossible to control if the main rotor slipstream recirculation vortex is blown into the tail rotor, especially if the vortex rotates in the same direction as the tail rotor (fig. 12.16a). Hovering in a

Fig. 12.15 *Tail rotor stall risk*

crosswind may create a vortex ring on the tail rotor (fig. 12.16b), and in quartering crosswinds the main rotor vortices may strike the tail rotor (fig. 12.16c).

Failure of the tail rotor drive shaft at high hover or climb-out power results in a rapidly accelerating spin unless the pilot immediately lowers the collective control and selects autorotation.

Loss of a tail rotor blade creates large centrifugal forces on the tail rotor installation, ripping out the transmission in a few seconds. Loss of the tail

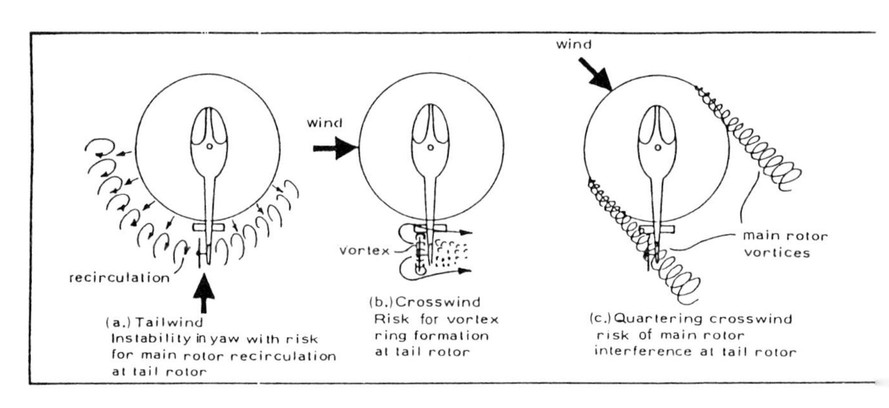

Fig. 12.16 *Directional stability and control problems while hovering in tailwinds and crosswinds*

rotor and transmission moves the helicopter's centre of gravity forward. As a result, the fuselage pitches nose down. If the pilot reacts by pulling the cyclic control rearward he tilts the rotorplane rearward and the blades cut through the upward moving tailboom. Thus the fin is cut off and the c.g. moves further forward. Recovery may be possible if the pilot immediately enters autorotation. The risk of uncontrollable spin and crash is large.

Tail rotor damage resulting in fatigue failures may be caused by tail rotor strikes in snow, water and twigs. Careful inspection and maintenance of the tail rotor is extremely important. Rotor strikes must be reported!

Hard Landings, Nose-over, Roll-over

The risk of misjudgement resulting in a hard landing increases with increasing descent angle, descent rate, payload and altitude. No vortex ring or blade contamination is required to make a landing flop if there is insufficient power available to stop the descent (fig. 12.17). This is especially true if the landing is made on a mountain peak or on ground with poor ground effect. In ground effect, induced drag is reduced and shallow-approach angle landings can be made where steep, high descent-rate approaches cannot be stopped (fig. 12.18). When landing in a headwind on a plateau watch out for the downsweep near the plateau's edge!

The c.g. of a helicopter is located fairly high above the skids. For this reason it is possible to nose-over a helicopter when landing on uneven ground at too high a speed. On rough ground the helicopter must be slowed down to zero ground speed before it is set down.

Sideways roll-over can develop on smooth surfaces if one skid contacts the ground with sufficient sideways speed (fig. 12.19).

Fig. 12.17 *Insufficient power for deceleration*

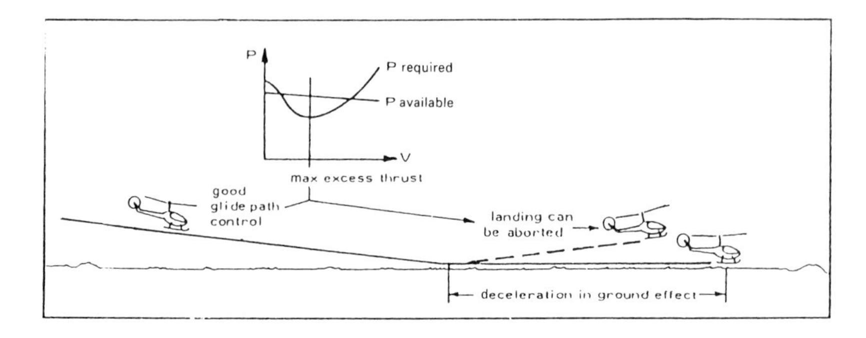

Fig. 12.18 *Landing in ground effect*

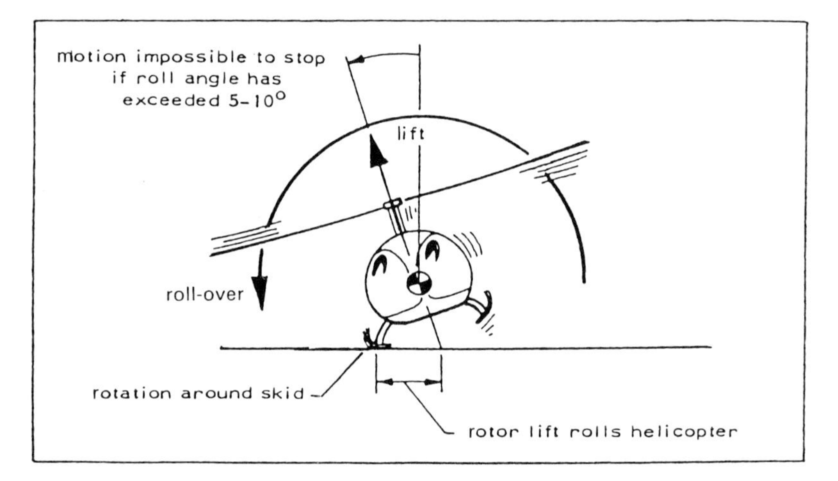

Fig. 12.19 *Dynamic roll-over*

The risks of sideways motion and dynamic roll-over increase in cross-winds, especially crosswind autorotation landings with forward speed and during take-offs from sloping ground.

Take-off in Ground Effect

Take-offs can be made in ground effect in conditions where the available power is insufficient for hover out of the ground effect. However, sudden loss of ground effect during acceleration to safe climbing speed may result in rapid loss of altitude and a crash-landing beyond the cliff edge unless there is sufficient altitude available for acceleration in a dive to safe flight speed (fig. 12.20).

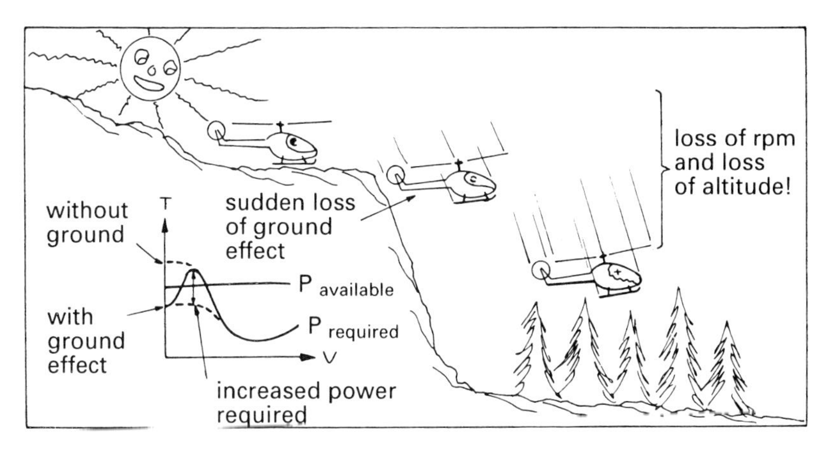

Fig. 12.20 *Sudden loss of ground effect*

Collision with Ground, Wires, Trees and Other Objects

Controlled flight into the ground is not only an I.F.R. problem. If height above the ground is not carefully checked during take-off acceleration or during an approach over a surface with poor references such as a mirror-smooth water surface or snow-covered ground the risks of flying into the ground are fairly large. Haze or blowing sand and snow above featureless surfaces add to the problem.

Wires often run along edges of fields surrounded by trees where they are difficult to see. Telephone and high-tension wires cross valleys and rivers and may be difficult to avoid if they suddenly appear beyond the river's bend and on the other side of a bridge you are 'jumping' over.

Thin support wires near structures and buildings and close to heavy high-tension wires may be very difficult to see. A little reconnaissance before landing pays off. Wire strikes are often fatal (fig. 12.21). The pilot of the helicopter in the picture did not check to see that the power lines had been strung further along the mountain than on his previous take-off.

Collision with objects during take-offs from and landings in confined places is probably one of the most common causes of helicopter accidents and incidents.

Behind these accidents/incidents is the certain belief that if a pilot has made a precision landing into and out of a 'tight squeeze' ten times all pilots can do it repeatedly. This belief does not consider the effects of unexpected distractions, recirculation disturbances caused by objects moved into the area and various types of wind disturbance such as hot air trapped in the clearing which rises suddenly when the helicopter approaches.

The 'confined space' type of problem exists even when operating from areas cleared in snow. The snow's edge serves as a recirculation deflector.

Fig. 12.21 *A fatal wire strike*

Asymmetrical recirculation tilts the rotor and makes the helicopter drift. If a skid 'bear-paw' hooks into the snow's edge when the pilot raises the collective control the helicopter rolls over. This happened during military winter operations. When the helicopter rolled, the rotor cut through the cabin and killed the mechanic. The costs of preventing the accident by doubling the area cleared would have been insignificant compared to the cost of the accident.

Loss of External References

When snow or dust is blown up from the ground during lift-off and recirculated into the rotor, the pilot suddenly finds himself inside a 'dome' with no external references (fig. 12.22). There is no time for transition to instrument flying since control is lost in a couple of seconds. The 'white-out' usually results in collision with the ground and roll-over. Watch out while flying in low clouds. It is possible to be trapped in seconds. Precautionary landing may then be too late.

In mountain terrain one valley may be filled with clouds while the next has clear weather. Flights around sharp bends may bring you into drifting clouds in seconds. Concentrate on attitude and altitude and make a quick 180° turn out of the mess.

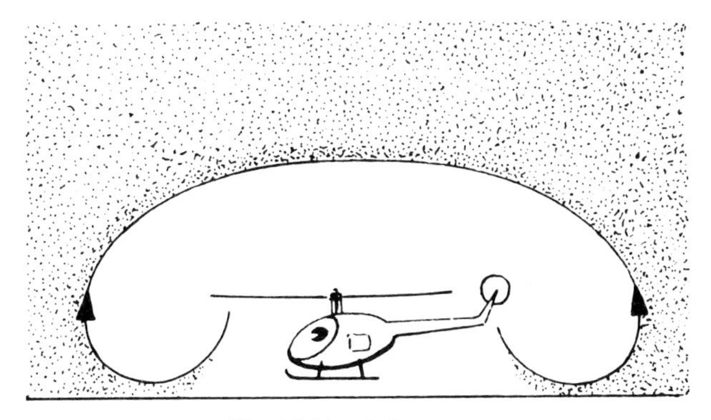

Fig. 12.22 *White-out*

Flights into Dangerous Weather and Winds

Flight into heavy turbulence always involves the risk of excessive blade flapping and of a rotor blade striking the fuselage. Landing in strong winds and close to updraughts near plateau edges sends blades 'sailing' above and below the flapping limits when the rotor slows down and the lift of the retreating blade approaches zero (fig. 12.23).

Lightning strikes on helicopters may delaminate composite blades (fig. 12.24) and burn rotating parts in transmissions. Fatigue cracks may develop from the burns.

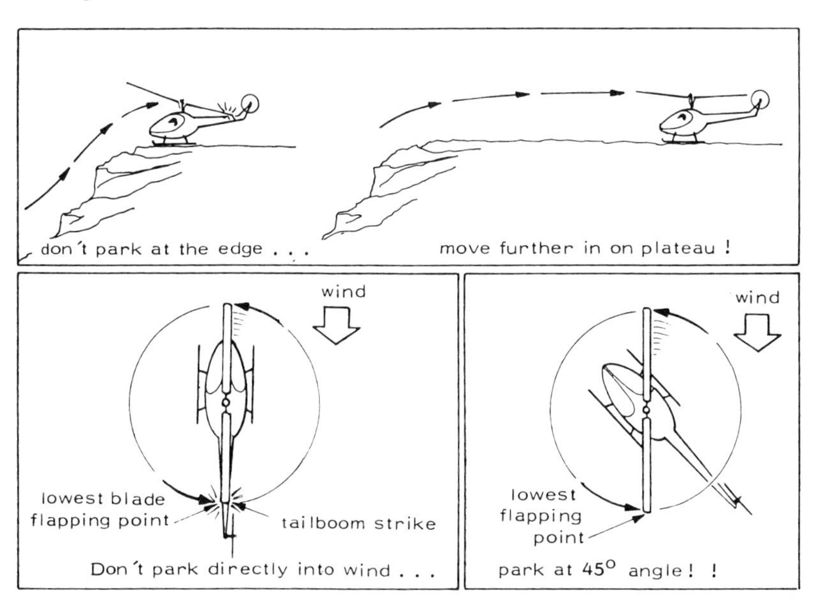

Fig. 12.23 *Parking in strong winds*

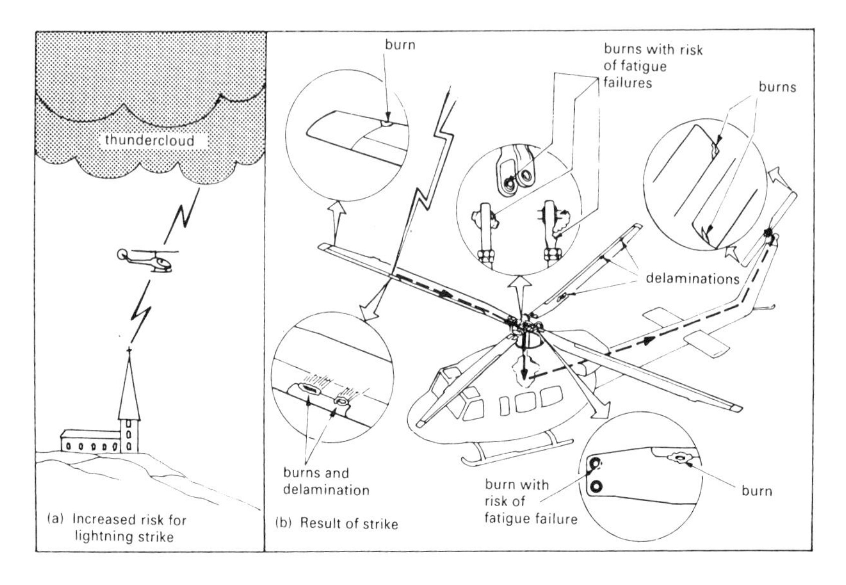

Fig. 12.24 *Results of lightning strike*

Fuel Exhaustion

Unless you are sure of two things, namely – 'I can always make 100% safe autorotations and I always fly over areas where safe autorotation is possible' – fuel your helicopter with ample reserves for the flight. Remember that fuel consumption is affected by headwinds, turbulence, rain, snow, icing, rotor blade contaminations and helicopter wear.

A pilot who, years ago, survived being shot down from roughly 65 000 ft altitude lost his life in an autorotation crash when his helicopter ran out of fuel.

By the way, remember that your available autorotation landing area increases more than twice as fast as your flight altitude above the ground, taking into consideration that some height is lost while establishing safe and steady autorotation.

Engine Failures, Rotor Failures, Airframe Failures

Engines, rotors and airframes may fail if helicopters fly into severe hail-storms or are struck by large birds.

Most failures between scheduled overhauls are, however, the result of excessive operational wear. Maintenance periods are not adjusted to suit the operational environment. The helicopter in fig. 12.25 crashed during an autorotation landing after a driveshaft fatigue failure. The helicopter had

Fig. 12.25 *Driveshaft failure*

been used for more than a year for forest fertilising operations which require extensive high-power flights for hover-loading of fertilisers and fast flights to the fertilising zones.

The helicopter was transferred from forest fertilising to geological ground tests in a mountain region. During the accident flight, the helicopter carried, in addition to the pilot, an observer and four geologists making ground soil tests. The four geologists were let off, took soil tests and picked up again continuously in accordance with an established test plan. During the three days preceding the accident, the pilot made more than 700 take-offs and landings. Helicopters are not designed for nearly continuous flight at maximum power without special maintenance.

Several heavy, used army helicopters crashed due to fatigue failures when transferred from military use to lumber slinging flights. The operators who bought the helicopters used the army maintenance schedules. However, the loads imposed on helicopters slinging heavy logs at high altitudes are much more severe than the loads in army service. Time between overhaul had to be cut down to one fifth of the time used by the army.

Helicopters used exclusively for high-speed cruise flights with full passenger loads in turbulent environments may wear out rotor systems in considerably shorter time than the same helicopters used for more mixed types of operation in gentler flight conditions. This is especially the case in salty atmospheres where salt contamination on the blades increases blade vibration and promotes corrosion which leads to accelerated crack propagation initiated in corrosion pits and finally to fatigue failure.

Hard landings induce whiplash loads on the tailboom. As a result, permanent deformations may occur in the tailboom attachment area. Cracks may develop in the deformed structure and fatigue failure may occur long before the helicopter is scheduled for overhaul (fig. 12.26). The risk is especially high if the helicopter is flown in a turbulent environment where in-flight *g* load variations result in repeated load variations on the tailboom and where landing in turbulence may induce repeated high loads both on the tailboom and the main rotor. Early crack detection is important in these cases since crack propagation may be large.

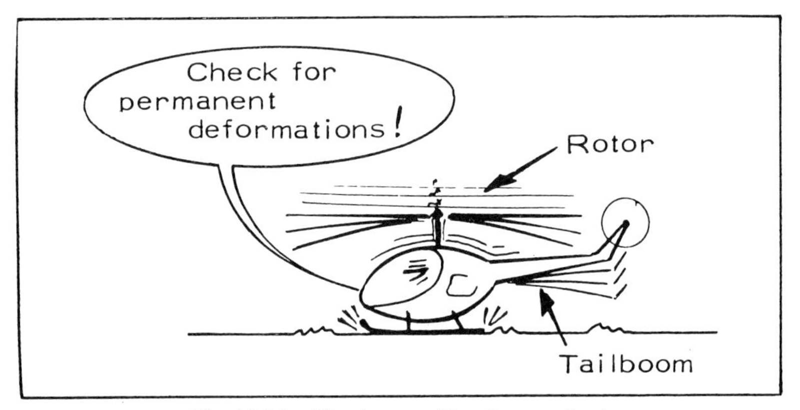

Fig. 12.26 *The danger of landing overloads*

The following case shows how fast fatigue failures may develop. The operator was flying bundles of plant for reforestation in a mountain region. Some bundles had to be watered to prevent the plants from withering and the bundles became too heavy to hover-lift. The company solved the problem by increasing the sling length so that the pilot could accelerate to sufficient forward speed for continued flight and climb before the bundles were jerked off the ground. The tailboom attachment failed after a few flights.

Sling-loading

Why is it so difficult to remember that unloaded slings (with nets) blow into the tail rotor in forward flight and that loads caught by bushes, trees, etc. pull the helicopter nose-first into the ground? There are repeated sling-loading accidents. Fig. 12.27 shows problems to remember during sling-loading operations.

The helicopter is a safe and versatile aircraft if you know how to handle it.

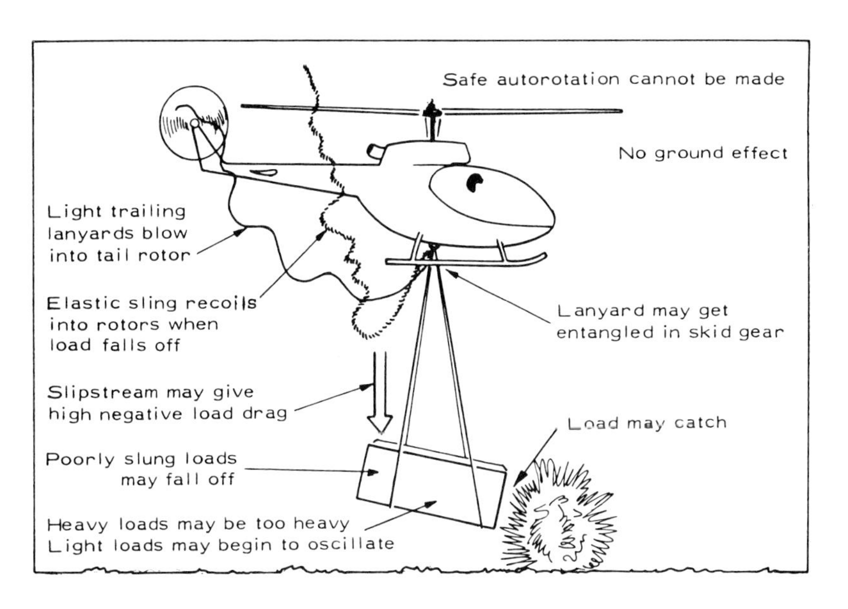

Fig. 12.27 *Sling-loading problems*

Chapter 13
Common Accident Causes

This chapter reviews twelve months of aircraft accidents, worldwide (1994, Airclaims, ref. I.S.A.S.I. Forum). The year 1994 was not a particularly bad year. The sort of accidents that happened during this year will, therefore, continue to happen during the years to come unless risk awareness is increased and accident-prevention work is improved. This should be possible since there is plenty of information available regarding the common accident causes.

Jet Transport Accidents

During 1994, thirty-one jet transports were lost in accidents. The majority of the accidents, nineteen (approximately 61%), happened during landing. Eight (approximately 26%) happened during take-off, two occurred *en route* and two on the ground.

The following types of aircraft were lost:

- two Tu-154Ms
- one A340-211
- one A310-308
- two A300-G22Rs
- one A330-321
- four Yak-40s
- one Il-86
- five Boeing 737s (various models)
- one Caravelle
- one Boeing 727-44F
- one DC-8-51F

- three F28s (M1000-4000-6000)
- one Tu-134
- one BAC 1-11
- one An-72
- two Boeing 707s
- three DC-9s (31, 31 and 32).

One of the *en route* accidents was a result of a midair collision and the second was caused by very incompetent handling of the situation in a 'glass cockpit'. The two aircraft lost while on ground caught fire.

The take-off/climb-out accidents included cases of flying into high ground during climb-out, take-off with a known technical problem resulting in fire and loss of control, inability to handle complex systems and system failure, nose-gear failure (possibly induced by previous hard landings) and one case of pitch-up and stall during touch-and-go flight training. The pitch-up could have been a result of rotation at too low a speed, wing leading edge contamination, which can make swept-back wings pitch up at lower than normal angles of attack, or a combination of both.

In nearly 37% of the landing accidents, the crews landed the aircraft at high speeds far down wet runways in reduced visibility instead of aborting the landings and making a new approach. Blown tyres, thunderstorms and typhoons were involved.

In 32% of the accidents, the crews collided with structures during approach, landed short or very hard, and sheared off the landing-gear. These accidents are signs of destabilised, poorly controlled approaches where executing missed approaches could have prevented accidents. In one case, the aircraft ran out of fuel on the way to the alternative airport in poor weather and in another the crew stalled the aircraft when pulling up into a strong tailwind during a discontinued approach. In one case, the landing-gear collapsed due to hydraulic system failure and in another the crew lost control during approach, possibly due to a runaway trim. Runaway trims can be very insidious since the crew does not necessarily notice anything before the autopilot disconnects, when the aircraft may react suddenly and violently to the faulty trim setting.

Finally, in one case the aircraft pitched up, climbed and stalled because the crew did not know how to handle a modern 'glass cockpit' system. Over the years we have seen several of these accidents where crews do not understand the systems or are so confident that the 'automatic flight system' handles the aircraft perfectly that they let the system fly the aircraft into the ground.

Looking back at the jet transport take-off and landing accidents we find that well-trained, attentive crews with high motivation and risk awareness could have prevented nearly all of them. There is a definite need for an improvement of cockpit resources management training for crews flying modern jets with complex electronic systems.

Regional and Executive Flights

During 1994 approximately 150 aircraft were lost. Most of these were smaller jets, turboprops and light twins with reciprocating engines.

The relative number of technical problems was much higher than for jet transports (26% compared to 13%) indicating less-efficient maintenance. Nearly 67% of the technical problems were engine failures. Since most of the engines were reciprocating this reflects the problem of maintaining the engines to the high standards required for commercial operations.

Other cases involved tyre failures during take-offs (three), one case of electrical fire and one of fire in the cargo hold. Finally, the undercarriage was retracted during landing in two cases and, in one case, a hydraulic system failure caused an accident.

When engines failed during take-off aircraft stalled in some cases and spun when the pilot (crew) attempted to return to the airport or did not manage to control a single engine failure on a light twin at low speeds. In very many cases, altitude could not be maintained on light to medium-size (DC-3) aircraft when one engine failed. The crews that crash-landed aircraft nearly straight ahead managed to survive.

Two aircraft were extensively damaged when they flew into hailstorms. In one case, the front windshield was knocked in.

One aircraft lost part of the empennage section in flight, possibly due to elevator flutter. Flutter may develop in high-speed flight on aircraft with play in the control system.

Approximately 10% of the accidents were caused by collisions with high ground *en route*. The evidence of poor navigation and follow-up of flight progress is clear.

Take-off and climb-out accidents accounted for 19% of the aircraft losses. In seven cases, aircraft stalled shortly after lift-off due to being overweight, too early a rotation, flaps being retracted too early or because of a steep turn immediately after take-off.

Two aircraft flew into the ocean when turning steeply after take-off at night (possible 'black hole' problem). Several collided with fences, trees or the ground after take-offs in I.M.C. or gusty winds. One aircraft veered off the runway, and two overran the end of the runway after aborting take-off below V_1. One aircraft had tyre failures, stopped on the runway but burned when fire spread from the undercarriage to the wing.

The majority of the non-technical accidents occurred during approach and landing (forty-one or approximately 27%).

The following accidents are typical.

- Ten aircraft flew into high ground on approach, collided with wires or trees or flew into water in haze ('black hole').
- Five aircraft landed short or sheared off their undercarriages.
- Four aircraft were extensively damaged in hard landings.
- There were two wing stalls due to icing and one dive into the ground due to tailplane ice.
- Three gear-up landings were made.
- Two aircraft stalled on approach.
- There were several accidents due to loss of control after touchdown, overruns after landing far down the runway, and landing next to the runway.
- In one case, fire broke out due to overheated brakes.

If we were to review where the fixed-wing aircraft accidents happen we would find that most occurred where educational systems are poor, safety training is poor, risk awareness low, management understanding of safety needs is poor, and the general respect for safety rules and regulations is low. Improved training could prevent many of the accidents.

Helicopter Accidents

During 1994, 183 helicopters crashed. There were a number of engine and control system failures but the majority of the accidents were caused by human error.

Nearly 16.5% (thirty) of the accidents happened during take-off and climb-out when pilots flew into power lines, trees, a crane, hills, water and snow surfaces, rolled over, spun or lost rotor r.p.m.

The collision accidents (50% of all take-off crashes) happened during day and night take-offs, and in both clear and cloudy weather. All could have been prevented with careful take-off and climb-out planning.

In six cases, helicopters rolled over or began to spin during take-off. The roll-overs occurred when one skid struck the ground in sideways motion due to crosswinds, recirculations, sloping terrain or a combination of both. In two cases, tail rotor driveshafts failed but in one case the pilot hovered at high take-off weight at high altitude on a warm day with only 90% torque which resulted in loss of rotor r.p.m. This reduced the tail rotor force which made the helicopter spin. The pilot increased torque to 100%. With high torque and reduced r.p.m. the spinning motion increased and the helicopter descended into the ground.

In nine cases, the helicopters lost rotor r.p.m. during take-off and descended into the ground. In two cases, the engines failed. In the remaining accidents there were no mechanical failures. Engines were overloaded and r.p.m. lost when pilots took off overweight at high altitudes on warm days. The engines flamed out when one pilot forgot to check for snow in the air intake. Forgotten preflight checks caused another flame-out.

Of thirty take-off accidents only four (13%) were caused by mechanical failure. The rest were preventable through better knowledge of the helicopter's limitations and improved preflight planning.

Nearly half of the accidents (46%) happened *en route.* Crews flew into power lines (thirteen), mountains, water surfaces, a frozen, snow-covered lake, and trees (thirty). The collisions happened in clear sunshine, at night and during flight over mist-covered terrain. The risks of continued low-level flights into deteriorating weather should be obvious. Loss of external references may result in loss of control in a few seconds. Power lines and wires may be nearly impossible to see but their whereabouts can usually be found on maps in preflight planning. Heights above water surfaces are difficult to judge, especially far from the shore in calm weather. Nevertheless, low-level flights are still made at high speeds over such surfaces.

Flights into severe turbulence in mountainous terrain caused two accidents. Turbulence and downdraughts may be difficult to see but not very difficult to predict. The leeward side of mountains is always dangerous in strong winds, and the wind speed at the entrance of a valley obviously increases and may become dangerously high as the valley narrows due to the venturi effect.

One accident was caused by flight into icing conditions; three crews ran out of fuel, autorotated, but crashed. These accidents were non-fatal but the helicopters were destroyed. Midair collisions caused two fatal accidents.

There were twenty cases of r.p.m. loss *en route* caused by engine failure, flame-out due to water in the fuel, transmission failure, and, possibly, exceeding maximum available power.

Loss of yaw control happened eight times. In some cases the driveshafts and transmissions failed but in other cases yaw control may have been lost during flights in tailwinds and into mountain vortices.

In one case a cabin door fell off and struck both the main and the tail rotors. Finally, three accidents were caused by main rotor control system failures.

Sling-loading caused seventeen accidents. In nine cases, power was lost when loads were picked up or delivered. In a couple of the cases the engines failed (high-power flights) but in at least one instance, power loss was a result of recirculation (vortex ring) developed while hovering among trees. In others, the loads may have been too heavy to lift. Tail rotor driveshafts and transmissions failed in four cases probably because maintenance frequencies were too low relative to the heavy-lift operations performed.

In three cases, a bucket and lifting lines hooked in objects on the ground. In two of the cases the lines pulled the helicopter into the ground and in one, the line, when pulled loose, recoiled into the main rotor which then failed. One sling slid over the bearpaw and rolled the helicopter when the helicopter climbed. Finally, there was a repeat performance of the accidents caused when lightly loaded slings and nets blew into the tail rotor as the helicopters moved forward.

During autorotation tests and training, five accidents happened when pilots through misjudgement flared too high or too late, and engines did not respond during recovery or flamed out. Only two accidents happened during spraying. One pilot flew into power lines and another lost r.p.m. in a steep turn after a spray run.

Hovering for various reasons caused nine accidents. A helicopter on a sightseeing flight hovering near a waterfall suddenly lost height and crashed into the stream below the fall. Friction between falling water and air creates downdraughts close to a waterfall and updraughts some distance away from it (in calm weather). A helicopter hovering close to the waterfall may drift from the updraught into the downdraught and descend into the ground. Hovering very close to the ground, especially in crosswinds or with rearward motion, involves the risk of skids striking the ground. In two cases the main rotor cut the tailboom.

High-altitude hover with one skid resting on the ground involves the risk of the main rotor striking rocks on a sloping mountainside or the risk of loss of control due to sudden gust disturbances. Two helicopters were lost for this reason. Another hovering 20 ft above a ridge suddenly spun due to local wind changes and struck rocks.

One helicopter hovering over a hot spring in a national park was engulfed in hot steam which reduced engine power and decreased visibility. The pilot lost control and crashed.

Prolonged hover over water resulted in rapidly increasing descent. A vortex ring may have been created.

During hover in a maintenance flight a canvas cover struck hydraulic lines and yaw control was lost.

Landing accidents accounted for 18.5% of the losses (34 out of 183). In eight cases, the crews landed hard in gusty crosswinds, reduced visibility, downdraughts and through plain misjudgement.

High-altitude landings accounted for another eight accidents. When helicopters slid off landing sites, rotors struck rocks, pilots lost control due to gusts, skids caught on rocks and tail rotors hit mountainsides.

Seven helicopters flew into water, the helipad, wires and airport runway thresholds in poor visibility and when blinded by the sun.

Two wooden platforms failed. In one case, a sunbed was blown up by the slipstream and recirculated into the main rotor when a helicopter landed on the deck of a large passenger ship. One helicopter rolled over when the pilot landed on the sloping shore of a river and two collided while landing.

Helicopters landed hard in two cases when engines failed during approach hover and in one instance when the tail rotor driveshaft failed. Passengers and crews survived. One crew heard an unusual noise during flight but continued to the destination where the helicopter hovered for ten minutes due to conflicting traffic before approaching for landing. Five metres above the ground the tail rotor drive failed. The helicopter spun, rolled over and was destroyed. Finally, in one case the helicopter landed hard and was destroyed when the passenger's knapsack caught on the throttle and pulled it to idle.

Concluding Remarks

Lack of training, lack of understanding of aircraft limitations and lack of risk awareness cause the majority of aircraft accidents from heavy transport operations to private pleasure flights. The accidents cost billions of dollars each year. Even small improvements could reduce the cost by millions, make flying safer and more pleasant, and eliminate much unnecessary human suffering.

Chapter 14

Ten Easy Do-It-Yourself Accidents for Private Pilots

The purpose of this chapter is to ridicule, not people but accident causes. Some readers may think this is funny and for this reason remember the risks. Others may become whooping angry and for this reason remember what they read. In both cases accidents are prevented. This is the purpose. Here are ten easy ways to have accidents.

RULE NUMBER ONE

"GET INTO THE SUPERMAN MOOD"

The basic rule In order to get into the right state of mind you must convince yourself that you are the tough master of all aircraft. You can kick those rudder pedals and pull that stick to the admiration of everybody. Theoretical nonsense, like aerodynamics, meteorology and human limitations to flight, is for the small birds. You are the eagle. You can handle all situations.

RULE NUMBER TWO

"WHO THE HECK NEEDS A PREFLIGHT CHECK"

The preflight rule Avoid all preflight checks. This can lead you into the most exciting situations. Always assume that:

- there are no fatigue cracks in the propeller
- there are no cuts in the tyres
- the cut or bulge in the canvas, or wrinkle in the skin, does not indicate that you may have structural damage
- the wet spots on the ground do not indicate fuel or oil leakage
- the control surfaces move freely (and in the right direction)
- the aircraft has sufficient fuel for the flight
- the fuel meter is always correct
- there are no insects in the pitot tube or in the fuel vent
- and so on . . .

Now you have the right setting for an accident.

RULE NUMBER THREE

"ENGINE STOPS MAKE FOR SHORT RANGE HOPS"

The take-off rule If you don't have much time to spare you should try something that can get you into immediate trouble, like this: no magneto checks, no engine run-up and no check-list reading before take-off. Just go zoom man!

Another interesting method has to do with the centre of gravity. Put some really stalwart fellows in the rear seats and add some heavy luggage in the rear compartment for good measure. You can't dream of the hair-raising control problems you will run into, especially if you have the trim at maximum nose-up.

Or, if you find these suggestions too crude, why don't you try a more subtle and refined combination of small factors. Then you can convince yourself, if you survive, that it was not your fault. Try combining the following.

- Select a medium altitude airport on a warm day.
- Make sure that it is a reasonably short field with a slightly soddy surface.
- Add an extra friend (your friends will enjoy this), just to make sure that you are slightly overloaded.
- Don't forget to add an 'unimportant' tailwind and a slight up-slope. That should make the day perfect.

- Ready? Go, go! Aim for the trees at the end of the field.

RULE NUMBER FOUR

"HEADWIND OR NO, FAR AWAY I SHALL GO"

The range rule In preparing to fail to reach your destination, here are a few good pointers.

- Meteorology is an exact science. The headwind will remain as reported. It will never increase.
- Fuel mixture has no effect on range.
- Flight altitude has no effect on range.
- Weight has no effect on range.
- The old crate you are flying has the same drag it had when it was new, especially if you carry external loads.
- The conditions at your destination will be ideal. You can land immediately and will need no alternative.

The interesting part about this type of flight is that you really don't know where you are going to end up for the next few hours, days, weeks . . . With a bit of luck you may become a new Tarzan or Robinson Crusoe.

RULE NUMBER FIVE

"LOSS OF SIGHT IS A FEARFUL DELIGHT"

The blind-flying rule Don't destroy the day by checking V.F.R. conditions along your intended route. Remember that 'every cloud has a silver lining'. Sing boy, sing, it will clear up in a moment and you have to get home anyway. In this case, the following advice for certain failure can be given.

- Never turn round.
- Never make a precautionary landing.
- Never believe the instruments. Your sense of balance is superior even when you cannot see. The fluid flow in *your* ear labyrinth can never give *you* false impressions.
- It is only nonsense that you must be on instruments in good time before entering a cloud. You don't lose control in a few seconds. You have plenty of time.
- There are no mountains, tall buildings or structures in the clouds.

The thrilling thing about going into a cloud without being equipped or prepared for it is that you really don't know where you will come out. Inverted spin from the cloud base is a good thriller. Beat that one for excitement if you can.

RULE NUMBER SIX

"STALL AND SPIN CAN DIG YOU IN"

The manoeuvre rule This accident can be practised nearly anywhere, but there are a few favourite spots such as above your girlfriend's country house, above your home to show Ma and Pa, or over a nudist colony where tough turning can keep your object of interest in view all the time. Some good suggestions to help you on your way are given below.

- Stalling speed is not affected by weight.
- Stalling speed is not affected by load factor (if you don't know what a load factor is, you are disqualified from our fun game since you will flunk without help).
- Wing contamination has no effect on stalling speed.
- Gusts have no effect on stalling speed.

With 'a little bit of blooming luck' you may crash through a barn roof and end up in the hay with the farmer's daughter (it has been done) or fall flat on the nudy-beach and have a delightful sight before you kick off.

P.S. If you don't like the above suggestions, try to cut one engine of a twin near the stalling speed. That should spin you, if nothing else does.

RULE NUMBER SEVEN

"A WINTRY WING IS AN EXCITING THING"

The winter rule 'Jingle bells, jingle bells, jingle all the way . . .' got your sky-sleigh ready? Have you left it outside during the night? Is it really good and frosty? Then you are all set to go for a good and spooky experience. Now, be sure not to brush off anything and, if you like, make the experience a little more spicy by neglecting all checks of frozen controls, and please, no pre-heating of anything before you go. Here are some famous last words that you may be remembered for:

- 'frost and snow blows off the wing and cannot, therefore, increase the stalling speed'
- 'that thin layer of ice left on the wing after the snow has been brushed off has no effect on take-off performance'.

Remarks like these indicate that you are a tough, winter pilot who cares nothing for theoretical nonsense like friction, boundary layers and wake drag. You are prepared to offer nearly all of your regular margin to stall for a quick departure.

Should winter conditions not be available at take-off, you can always go and look for some icing conditions in the clouds. That gives you another chance of increasing the stalling speed by up to 30%. Having found the ice, then the following is good negative advice.

- Don't change altitude, the icing conditions may disappear.
- Make a long and shallow approach so that you are sure to pick up plenty of ice before landing.

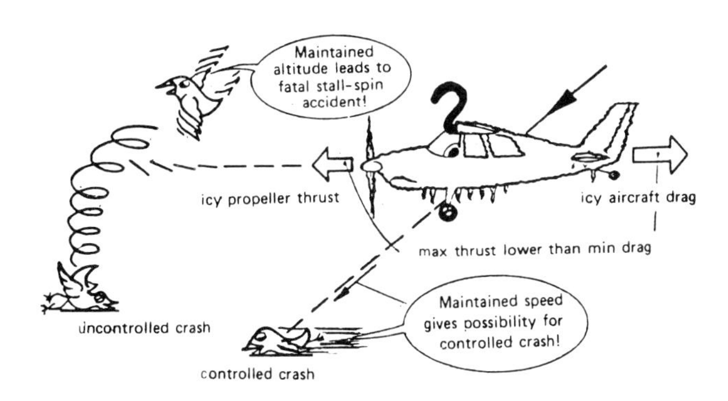

- Don't try to maintain flying speed, but maintain altitude until you get that shuddering feeling of stall. Of course, if you don't want to spin in you could maintain flying speed and make a controlled crash. (This part of the advice is not so negative, bad conscience suddenly struck me).

Now, let us say farewell to winter with this little rhyme:

Winter flight is such delight,
Frosty wing is the he-man's thing.

RULE NUMBER EIGHT

"MOUNTAIN FLIGHT IN SPOTS SO TIGHT"

The mountain rule 'Ah, the mountains, the lovely sights.' Sing a song to praise them: 'Cruising up the river, up towards the falls, narrowing the valley width, ahead are rocky walls . . . da dum, da dum, da dumda . . .' Now you have your chance to show some really exciting 'getting out of it' airmanship. What shall it be, pull-up with a top roll and a stall, or a half-roll and a dive into the ground? Or, shall we just settle for a 'precautionary' crash in the rocky river?

In preparing for this type of accident, you must convince yourself that:

- your aircraft can turn on a coin
- it can outclimb all types of rising ground, especially at high altitudes
- there are no updraughts, downdraughts, rotating flows or gusts in the mountains

- the wind direction never changes; if you made it over the ridge yesterday you will make it even today.

Good luck in the mountains!

P.S. Don't bring any survival equipment; it is much more interesting to try the primitive life of your forefathers (those way, way back).

The see-and-avoid rule If you have failed to follow any of the destructive advice I have given so far, you could always try to touch wings with someone. Get yourself into a busy terminal area with plenty of choices and don't tell anyone you're there. Go for a big surprise! Now, the assumptions and golden rules for success in this case are:

- don't listen to A.T.C.
- don't look round, everyone sees you
- there is plenty of time to avoid collision once you have seen each other

RULE NUMBER NINE

"MIDAIR COLLISION SPELLS THE END OF THE MISSION"

- there are no aircraft in the glaring sunset and none camouflaged by city lights
- no transport will come zooming up through the cloud layer below or wooshing down from the clouds above
- the transport crew has nothing to do, they hang out through the window just looking for little important you. All check lists have been abolished and replaced by faithful computers.

By the way, midair collision may be too drastic for excitement, so instead of hitting someone, why don't you just try to get into the trailing vortex of a heavy jet. That should give you an unforgettable ride, providing it does not tear off your wings.

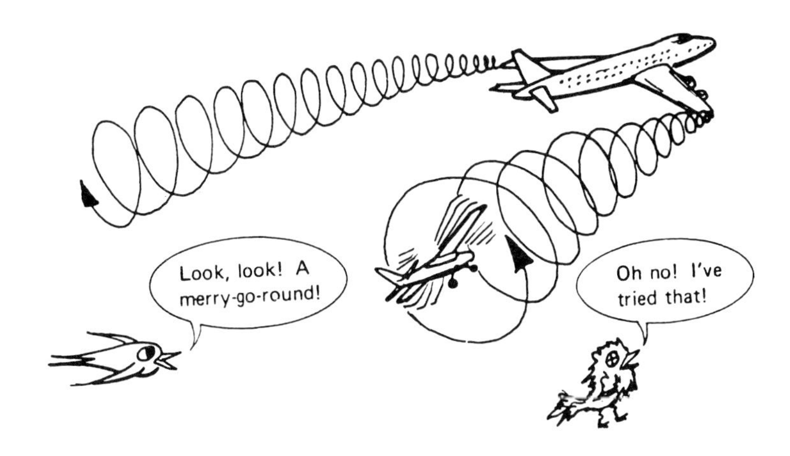

RULE NUMBER TEN

"JUST DROP DOWN AND LAND WITHOUT ANY PLAN"

The landing rule Here the possibilities are really tremendous. This is one of the favourite accident phases of flight. Select your faulty assumption from the list below.

- That nice, steady headwind will never change to a tailwind at touch-down.
- There will be no gusts or rotors due to terrain and buildings.
- You can easily see through the shallow fog covering the ground.

- You will soon break through the overcast and the airport will be right in front of you.
- 'Braking action good' means that the runway is not slippery so you may as well land down-slope in a tailwind.

By the way, by now you may have caught on, so why don't you fill in a few of your own favourites.

. . . .

. . . .

. . . .

Good luck!

Well, this was a short repetition of things that have been said before. I hope you remember the problems and avoid them. Flying is safe. All you need is sound sense and risk awareness.

Index